W9-AFH-991

Direct Digital Controls for HVAC Systems

Other HVAC Books From McGraw-Hill

Direct Digital Controls for HVAC Systems

Thomas B. Hartman, P.E.
The Hartman Company
Seattle, Washington

McGraw-Hill, Inc.
New York   San Francisco   Washington, D.C.   Auckland   Bogotá
Caracas   Lisbon   London   Madrid   Mexico City   Milan
Montreal   New Delhi   San Juan   Singapore
Sydney   Tokyo   Toronto

Library of Congress Cataloging-in-Publication Data

Hartman, Thomas B.
    Direct digital controls for HVAC systems / Thomas B. Hartman
        p.   cm.
    Includes index.
    ISBN 0-07-026977-7
    1.Heating—Control.   2. Ventilation—Control.   3. Air
conditioning—Control.   4. Digital control systems.   I. Title.
TH7226. H38   1993
697—dc20   92-20517
                                                CIP

This book is printed on recycled, acid-free paper containing a minimum of 50% recycled deinked fiber.

ISBN 0-07-026977-7

The sponsoring editor for this book was Robert W. Hauserman, the editing supervisor was Joseph Bertuna, and the production supervisor was Donald F. Schmidt. It was typeset in Palatino by the McGraw-Hill Professional Book Group composition unit.

Printed and bound by R. R. Donnelley & Sons Company.

Contents

Contents

Preface

Since the early 1970s when digital computers were introduced to heating, ventilation, and air-conditioning (HVAC) control applications, the transition to this new technology has been burdened by the legacy of pneumatic controls. A few in the industry saw that digital HVAC controls offered the promise of a clean slate for designers and building operators insofar as energy and comfort performance were concerned. But an overwhelming majority of the industry, especially controls manufacturers, sought configurations of digital controls that operated as closely as possible to the pneumatic controls they were to replace or supplement.

In the more than twenty years that have passed since direct digital control (DDC) systems have first been applied, little in the industry has changed. Most manufacturers are not reluctant to characterize their development effort as an evolutionary process. Unfortunately, the pace of this development for many DDC products has actually been slowing due to complacency within the industry and the complexity of products now being offered. The result is that the gap between user's expectations and system performance is beginning to increase for the first time in a long while.

The reasons for this widening gap between control system performance and user expectations are twofold. First, the computer revolution is spawning a new breed of building operators who are not nearly as sympathetic as old-time pneumatic technicians with the complexity required to perform simple control functions in buildings today. Second, users are frustrated by the inability of HVAC systems to meet the rising expectations of their tenants insofar as comfort and air quality are concerned. More sophisticated users are further discouraged by the energy and operating inefficiencies of many systems as well as the lack of individual control capabilities. While some of these problems relate to deficiencies of mechanical systems as a whole, a far larger number are correctly perceived to be directly attributable to unnecessary limitations of the controls.

There is today a growing concern within the HVAC industry that it could soon become a target for criticism, undesirable regulation, and litigation if building comfort systems do not begin doing a better job of meeting the building occupants' rising expectations for comfort and environmental quality. This concern is well founded in light of the fact that many systems being implemented today do not comply with ASHRAE standards regarding comfort and outside-air ventilation.

But meeting these existing and emerging standards of building operation is not difficult. Our firm uses the term *high-performance* to delineate HVAC systems with equipment that obtain higher levels of performance never before possible from systems with equipment that obtain past levels of performance. For example, *dynamic control* (i.e., anticipatory time-changing control) is a high-performance replacement for steady-state control, which characterizes the operation of most past and current HVAC systems. *Integrated control* is a high-performance replacement for unitary control, which underlies pneumatic control design. Each of these operating principles can have enormously positive effects on building operation, and at the same time they offer improved energy and operating efficiencies. There are many more examples of high-performance replacement strategies that are waiting for forward-thinking designers and operators to identify and exploit.

The HVAC industry can be "reinvigorated" by efforts to develop and better understand the benefits of high-performance HVAC systems. As always, effective high-performance designs will drive the industry to develop products that meet the functional requirements of those designs. This book is intended for those in the industry who are interested in developing high-performance applications for DDC systems. The material is aimed at all who have responsibility for good HVAC system operations whether they work as consulting engineers, building operators, or DDC system manufacturers or contractors.

Thomas B. Hartman

Direct Digital
Controls
for HVAC
Systems

1

The Challenges and Benefits of High-Performance Systems

The opportunities for improved performance of HVAC systems that stem from the development of high-performance *direct digital controls* (DDCs) are rarely fully exploited in building mechanical system designs. A majority of design work in the building construction industry is derived from rules of thumb rather than rigorous analysis. Concepts that break established rules of design and operation are still considered risky. Designs utilizing high-performance controls require the designer to rethink how systems can be most efficiently configured, and this thinking challenges the rules of thumb that are widely applied. To see the benefits of more rigorous design approaches and to challenge our acceptance of traditional design practice, consider the operation of a variable-flow chilled water loop.

Variable-Flow Chilled Water Loop

Figure 1-1 is a schematic of a chilled water system employing variable flow on the secondary (load) circuit. The loads are cooling coils in the building's various air systems. In this case, the designer is confronted

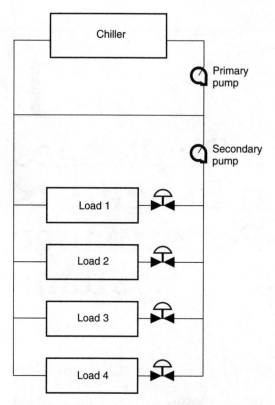

Figure 1-1. Primary/secondary chilled water loop.

with a chiller that serves an extensive piping array with multiple chilled water coils—a reasonably common occurrence. Assume that all the cooling loads vary fairly uniformly with outdoor conditions and that the coils operate at low-load conditions much of the time as is typical of HVAC cooling systems. To reduce pumping energy, the designer decides to incorporate a variable-speed pump on the secondary loop. To decouple the constant chilled water flow requirement of the chiller evaporator from the variable flow through the loads, the designer decides upon a two-pump arrangement with two-way valves for each load and a variable-speed drive to operate the secondary chilled water pump. The primary chilled water pump is a low-head, low-power design that operates continuously with the chiller.

To design the secondary loop, traditional design manuals suggest sizing control valves for a full-flow pressure drop of at least 30 per-

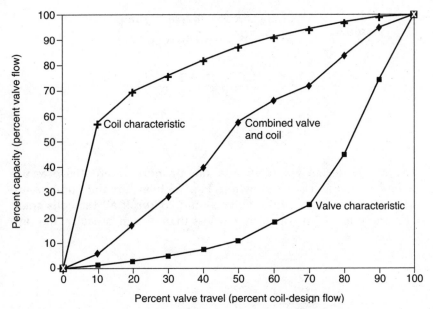

Figure 1-2. Design of chilled-water valve and coil combinations.

cent of the total system drop. This factor is often called *valve authority*, and rules of thumb dictate that it should be equal to or greater than the pressure drop through the load, in this case the cooling coils. The substantial pressure drop through each valve is an effort to attain some measure of linearity between valve action and the cooling effect delivered to the load. The high pressure drop also serves to isolate the operation of each valve/coil arrangement from other. Each valve is matched to the coil it operates to provide as near a linear capacity response to valve actuator travel as possible. Figure 1-2 shows how a valve and coil might be matched under traditional design rules to provide an overall linear relations between the valve movement and the cooling capacity delivered.

Depending on the physical location and piping lengths, an arrangement of 20-ft water-column drop across the loads, 30-ft drop across the valves, and perhaps 40-ft head drop across the secondary piping at full-flow conditions would be typical for a design using published rules of thumb. Assuming a total full-flow requirement of all the load coils of 1000 gal/min, the theoretical pumping horsepower at full flow can be calculated as follows:

$$\text{Pump hp} = \frac{(\text{gal}/\text{min}) \times 8.35 \text{ lb}/\text{gal} \times \text{pump head (ft)}}{33,000 \text{ ft} \cdot \text{lb}/(\text{min} \cdot \text{hp})}$$

$$= \frac{1000 \times 8.35 \times (30 + 20 + 40)}{33,000}$$

$$= 22.8 \text{ hp}$$

A traditional design would involve one or more differential pressure transducers in the loop as shown in Fig. 1-3 to ensure the pump maintains a 50-ft head across the valve(s) and coil(s). If all the coils spent large amounts of time operating at less than design capacity, the pip-

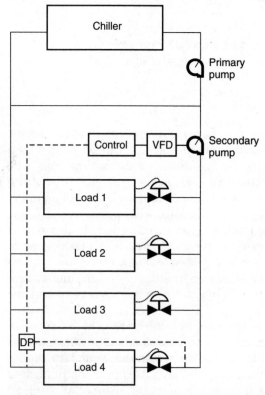

Figure 1-3. Primary/secondary chilled water loop with differential pressure (DP) sensor and controller.

ing head loss would drop with the square of the water flow decrease. The pressure drop through the coil would similarly decrease, but the savings would be excluded because the controls will keep a constant pressure across the valve/coils—the pressure drop across the valves will increase as the flow decreases. If reasonably even load profiles for all coils are assumed, the pipe head drop will drop with the square of the flow. At 75 percent flow, the piping head loss is

$$40 \times 0.75^2 = 22.5\text{-ft water column}$$

Using this new pressure-drop figure for the piping, we see that the horsepower requirement at 75 percent flow is

$$\frac{(1000 \times 0.75) \times 8.35 \times (50 + 22.5)}{33,000} = 13.8 \text{ hp}$$

At 50 percent of design flow, the piping head loss is

$$40 \times 0.5^2 = 10 \text{ ft}$$

The horsepower requirement at 50 percent flow is

$$\frac{(1000 \times 0.5) \times 8.35 \times (50 + 10)}{33,000} = 7.5 \text{ hp}$$

The designer would them simulate or estimate the amount of time the pump and coils would spend at various conditions and calculate the power requirements as above. If the system operates long hours at low loads, the savings will be substantial.

High-Performance Design Issues

The above analysis, simplified though it is, shows how substantial energy-use reductions are possible by the application of variable flow to typical hydronic loops in HVAC systems. Designers with a good understanding of high-performance direct digital controls (DDCs) should ask themselves why they allow their designs to be limited by traditional design methodology. Design teams often lose track of the *reasons* for the rules of thumb they apply so regularly. When this happens, designers run the risk of applying outdated design techniques— an important reason why many designs fail to meet expected levels of performance. What follows are the basic issues that underlie the development and use of rules regarding valve selection and a discussion of the reasons why the availability of high-performance controls should cause designers to rethink the applicability of those rules.

Linear Response Between Valve Travel and Coil Capacity

Design manuals stress the need to select valve/coil combinations for which equal increments in valve position will effect equal increments in heat transfer of the coil throughout the stroke of the valve actuator. Figure 1-2 shows how traditional design practice seeks to "linearize" the overall performance of the valve and coil. This is done because under traditional control it is assumed that the valve will be operated by a controller with a fixed proportional gain. Although this design principle is widely employed, it is not often useful to modern HVAC applications. For example, in *variable-air-volume* (VAV) applications the variable-airflow characteristics on the load side of cooling coils act to dynamically change the heat-transfer characteristics of the valve/coil arrangement as airflows change and destroy the linear performance that has been based on a constant load-side airflow.

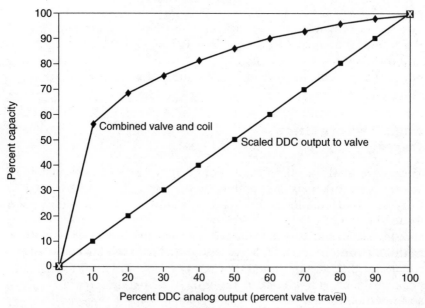

Figure 1-4. Use of scaling to linearize control.

High-performance DDC systems permit designers far greater flexibility in their design of modulating control systems without relying on assumptions that may not be realistic. Figure 1-4 shows a valve-and-coil combination that does not provide a linear response of valve position to coil capacity. However, many high-performance DDC systems permit scaling factors to be applied to the analog outputs. This permits an inherently nonlinear device to respond in a linear fashion to signals from the DDC system. In this example, the valve-and-coil combination provides about 70 percent of the design cooling capacity at about 20 percent valve travel. The DDC output to the valve can be adjusted via a scaling table to position the valve at 20 percent travel at a 70 percent output signal from the DDC system. The scaling factor allows standard *proportional, integral, and derivative* (PID) control to operate the valve effectively because of a "software" linearization of the valve/coil combination.

However, the chilled water flow versus heat-transfer performance assumed for Fig. 1-4 is valid only for unvarying load-side flows and inlet temperatures and for constant chilled water supply temperatures. Whether inherent in the system design or for optimization reasons, rarely in real HVAC applications do these other variables remain constant as control loops operate. This issue of linear control has therefore been only weakly resolved in the past by attempting to linearize components at one set of system conditions. Obtaining good control over wide ranges of system conditions can be achieved far more completely and effectively with high-performance DDC systems. The proportional, integral, and derivative gains can be tied to algorithms that adjust their values as the variables such as load-side flows and temperatures and chilled water temperature change. Even more impressive is the emergence of self-tuning controllers. These controllers continually reestablish the various gains associated with a control loop to provide more accurate control without hunting. The benefits of self-tuning are especially important because variables beyond the immediate control loop can have profound and widely varying effects on each control loop. Self-tuning features are becoming widely available with high-performance DDC systems and are enormously effective in adjusting control loops to continue stable operation as other system variables change.

Controllability

Selecting equipment for linear response or to isolate one control loop from another is not an overriding consideration for designers schooled in the application of high-performance DDC. However, this does not mean designers can be imprecise in their designs or in the

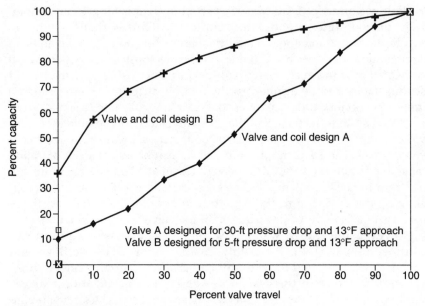

Figure 1-5. Design of valve and coil combinations.

selection of control-loop components. The issue of controllability will continue to play a prominent role in both the design of systems and the selection of individual components. Controllability remains largely a sizing issue. If a valve is oversized for given conditions such that the smallest increment possible from the control loop will substantially overshoot the desired control conditions, the loop has become uncontrollable. This is a problem that typically emerges during periods of low load. By fully understanding the issue of controllability and applying high-performance DDC systems correctly, designers can solve such problems and at the same time vastly improve the efficiency and performance of these systems.

Selecting a control valve with a lower pressure drop will reduce the pumping power required to meet the load conditions. Traditional practice strongly condemns the idea of employing large valves with lower pressure drops because of the nonlinear response and the lack of controllability at low loads. Figure 1-5 illustrates the dilemma. The valve/coil combination with valve A may be selected according to traditional design practice because it is reasonably controllable down to low-capacity requirements. The vertical axis intercept represents the smallest incremental cooling transfer possible as the valve is cracked

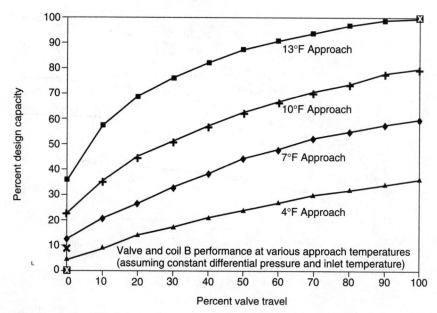

Figure 1-6. Heat transfer versus valve travel with variable chilled-water approach temperatures.

open. Note that it is small—only about 10 percent of the design maximum cooling rate. The coil combination with valve B has a much lower pressure drop because valve B is a larger valve. While valve/coil combination B would require less pumping power, the y axis intercept is much higher than that for combination A. Traditional design criteria typically declare valve B unsuitable for the application because it is uncontrollable at lower loads and because the valve position–cooling capacity relationship is nonlinear. But when it is integrated with a high-performance control system that can adjust both the chilled water temperature and the loop pressure, will linearity and controllability of the B combination really be a problem?

System Dynamics Considerations

To see how this question can be answered, consider the graphs in Figs. 1-6 and 1-7. Figure 1-6 shows the operation curves for the B valve/coil combination at a number of different approach (chilled water supply less air temperature leaving coil) temperature condi-

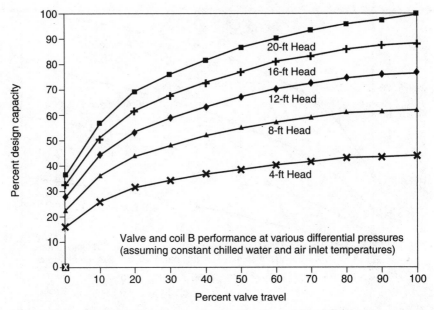

Figure 1-7. Heat transfer versus valve travel with variable differential pressure.

tions. It is clear that by increasing the chilled water temperature relative to the leaving air temperature, the controllability at low loads is
markedly improved. Similarly, Fig. 1-7 illustrates that the decrease in
pressure across the valve/coil combination also improves the controllability at low loads.

Designers can use these relationships to substantially reduce the
problem of controllability. At periods of uniform low loads, the DDC
system can reduce the head pressure across a valve and increase the
chilled water temperature to improve controllability. If all valves on a
common chilled water loop experience similar decreases in load con-
currently, this parameter adjustment is a great help in improving controllability at low loads.

It is apparent from Figs. 1-6 and 1-7 that larger rangeability and
low-load controllability are achieved by controlling the chilled water
temperature for load adjustment. Raising the chilled water temperature provides a bonus of chiller efficiency increases, but the chilled
water adjustment reduces pumping savings because a higher chilled
water temperature increases the water flow necessary to meet loads.
Additionally, under certain circumstances, dehumidification requirements may limit the permissible chilled water adjustment.

A high-performance DDC system can make the B valve/coil combination perform very well by increasing the chilled water temperature and decreasing the differential pressure of the loop as the loads on the coils decrease. Obviously, it is wise to keep the chilled water temperature adjustment at levels that permit flow reductions in order to maximize pump power savings, but adjusting the chilled water temperature upward offers savings, too. A high-performance DDC system can be operated to maximize the savings at all times. A DDC system must be chosen that offers the use of advanced algorithms, to adjust the chilled water temperature and head pressure according to load, and provides the ability to self-tune modulating loops. With this level of control, it is clear that valve/coil combination B can be operated effectively over a range from the design maximum cooling capacity all the way down to very low loads—generally much better than valve/coil combination A could be operated at fixed pressure and chilled water temperature conditions with lower performance control. The question remaining is how the system pressure and water temperature changes will affect the other valve/coil combinations on the cooling circuit. The assumption for the example is that all coils will experience similar loads, as is often the case with HVAC systems. However, the designer must be very careful to ensure that this is actually the case. If one of the coils serves an interior computer room that has high constant year-round cooling loads, or coils serve different perimeter zones of a building that are subject to high solar gains on the south side, those unusual zones may have to be specially accommodated by either a booster pump or a separate chilled water circuit, as shown in Figs. 1-8 and 1-9.

In Fig. 1-8, a small booster pump is added to increase the differential pressure for load 2 which the designer has determined will not fall as quickly as the others on the loop. In Fig. 1-9, two entirely separate loops have been configured to permit the separation of loads into groups that will have similar part-load patterns. The configuration in Fig. 1-9 may be cost-effective if the load groupings are in different locations and do not require extensive additional piping.

High-Performance
Chilled Water Control
Program Outline

Based on the above design data the designer may develop a high-performance control program for the loop with several basic elements as follows:

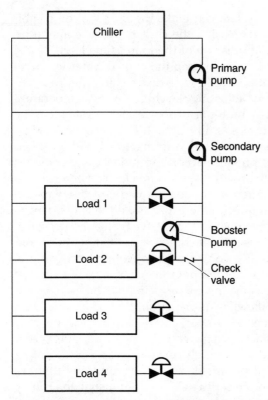

Figure 1-8. Primary/secondary chilled water
loop with booster pump on one load.

1. The pump speed will be decreased, and the chilled water temperature will be increased to ensure that at least one of the control
   valves serving the loads served by the loop is fully opened at all
   times during operation. The system will further ensure that all
   loads are met at all times.

2. In moving from a full-load to part-load situation, the pump speed
   will first be reduced to 90 percent of maximum speed while the
   chilled water temperature remains constant. Thereafter, the chilled
   water temperature will be reduced along with pump speed reductions at a relative reduction ratio that maintains a minimum valve
   opening of 20 percent.

Note that the control program outline does not anticipate the need for
a loop differential pressure input.

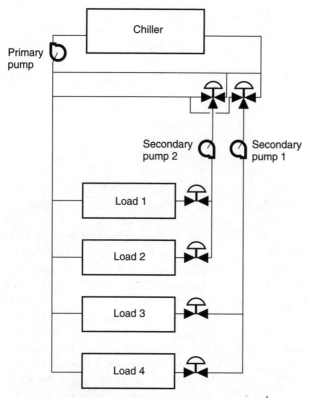

Figure 1-9. Primary/secondary chilled water loop with two secondary loops.

High-Performance Chilled Water Loop Design

Understanding that many traditional rules of thumb need not be applied to systems operating with high-performance DDC controls, consider how the chilled water loop of Fig. 1-1 may be designed for improved energy performance when it is operated by the above control program. With a high-performance control system, the control valves can be selected for lower pressure drops because the high-performance control program can be developed to handle the issues of linearity and controllability. Assume the same piping and coils are selected as in the earlier example except that the control valves are sized for a 5-ft pressure drop at design flow. Now the total head for the secondary loop at full-load conditions is the sum of the 20-ft drop across the loads, 5-ft drop across the valves, and 40-ft head drop across the secondary piping. At the full flow of 1000 gal/min, the theoretical full-load pumping horsepower can be calculated as

$$\text{Pump hp} = \frac{(\text{gal/min}) \times 8.35 \text{ lb/gal} \times \text{pump head (ft)}}{33,000 \text{ ft} \cdot \text{lb/(min} \cdot \text{hp)}}$$

$$= \frac{1000 \times 8.35 \times (5 + 20 + 40)}{33,000} = 16.4 \text{ hp}$$

Although this design requires larger valves to achieve a lower pressure drop at full flow, the pump, motor, and variable-frequency drive components are all only about 70 percent as large as required for the initial example, which means that the overall cost of the mechanical components of the design is likely to be the same or less than that of the initial design.

At lower loads, the pumping energy calculations are more complicated because both the chilled water temperature and the loop pressure will be adjusted by the high-performance DDC system to meet the specific requirements of the various loads. Raising the chilled water temperature raises the chilled water flow required to meet the loads, but also increases the efficiency of the cooling plant and results in additional energy-use reductions. Reducing the total loop pressure as the load decreases will enhance the energy savings beyond the savings from flow reduction. By using the performance graphs in Figs. 1-6 and 1-7, it can be assumed that at 75 percent average load the system operates at 11.5°F approach. At this operating point the flow would be

$$1000 \times \left(0.75 \times \frac{13}{11.5}\right) = 848 \text{ gal/min}$$

The piping and coil head losses are

$$40 \times \left(\frac{848}{1000}\right)^2 = 28.8 \text{ ft} \qquad \text{piping}$$

$$20 \times \left(\frac{848}{1000}\right)^2 = 14.4 \text{ ft} \qquad \text{coil}$$

The assumed head loss through the valves depends on the control strategy and the piping arrangement to be employed. Generally, the head loss through the valves can be assumed to decrease at the same ratio as the piping and coil head loss, but this is not always the case. To be conservative, it is assumed for this example that the valve head loss remains at 5 ft.

With these assumptions, the horsepower requirement at 75 percent flow is

$$\frac{848 \times 8.35 \times (5 + 14.4 + 28.8)}{33,000} = 10.3 \text{ hp}$$

This design clearly offers some additional part-load pumping savings over the original example, but more importantly it also offers savings from a 1.5°F increase in the chilled water temperature. Using a chiller savings of 0.02 kW/(ton[cdt]°F) energy reduction per degree increase in chilled water temperature yields an additional savings of more than 9 hp at the chiller.

At 50 percent of design flow, it can be assumed from Figs. 1-6 and 1-7 that the system operates a 10°F approach. At this operating point the flow is

$$1000 \times \left(\frac{0.5 \times 13}{10}\right) = 650 \text{ gal/min}$$

The piping and coil head losses are

$$40 \times \left(\frac{650}{1000}\right)^2 = 16.9 \text{ ft} \qquad \text{piping}$$

$$20 \times \left(\frac{650}{1000}\right)^2 = 8.5 \text{ ft} \qquad \text{coil}$$

To be conservative in advance of a certain piping layout, the head loss of the valves is assumed to remain constant at 5 ft.

With these assumptions the horsepower requirement at 50 percent flow is

$$\frac{650 \times 8.35 \times (5 + 8.5 + 16.9)}{33,000} = 5.0 \text{ hp}$$

As before, the part-load operation pumping savings are again overshadowed by the 3°F increase in the chilled water temperature. Using the chiller savings of 0.02 kW/(ton · °F) energy reduction per degree increase in chilled water temperature yields an additional savings of about 7 hp at the chiller.

Benefits of High-Performance Design

The above example provides results that are typical from applying high-performance DDC to HVAC systems. These results for mechanical designs which are carefully integrated with high-performance controls are expected:

1. Smaller (and sometimes simpler) energy conversion equipment sizes with accompanying savings that can reduce the system costs or be invested in higher-quality, longer-lasting components

2. Far lower total system energy use at *all* loads than what is possible with nonintegrated design and control strategies

3. Control precision at all loads that is superior to that of traditional control approaches

High-Performance DDC Chilled Water Loop Design

Thus far, this chilled water loop design has only anticipated high-performance direct digital controls. In the minds of many, the additional cost necessary for high-performance controls will force the cost of such designs beyond a reasonable budget. There is no question that the control system cost can be a factor in the feasibility of an approach, but no generalizations should be made before the exact control point requirements are fully considered. Depending on point densities and point types, the current cost associated with implementing high-performance controls is usually less than $300 per point when the control system is competitively acquired (discussion on procurement procedures that enhance competition occurs later in this book). At these costs, high-performance DDC options may cost less than pneumatic or nonintegrated electronic control-based alternates. The designer needs to consider that high-performance control layout can usually be configured with fewer instrumentation points than traditional control approaches. The intelligence of the high-performance DDC system can sometimes be used to replace instrumentation.

As an example, consider the chilled water loop. The traditional control of the secondary pump, as shown in schematic form in Fig. 1-3, requires one or more differential pressure (DP) sensors and a controller to operate the variable-frequency drive that sets the pump speed. Also required are control points to operate each valve. Assuming these are all chilled water coils in air systems, the instrumentation includes a discharge air sensor, valve actuator, and controller. A high-performance DDC configuration can be simpler because the system uses the output signals to each valve along with the air temperature conditions to determine if the loads are being met with the current system settings. Assuming the DDC system is integrated into all HVAC components, the chilled water temperature is already included as a point for the control of the chiller plant.

The next question involves the location of the differential pressure sensor or sensors. Is a differential pressure sensor really required for control of this loop? The control program outline did not use differential pressure as an input. All the information necessary to operate this sequence is available to the DDC system by simply interacting with the local valve control loops. The result is a high-performance control scheme that is actually simpler and costs less than the traditional controls that it replaces and outperforms.

Points to Remember

To fully exploit the benefits of high-performance DDC systems, designers must "engineer" designs before solutions are selected. Rules of thumb have no place in a high-performance design process. The benefits in improved energy performance, and in many instances lower first costs, are compelling reasons for designers to acquire the tools and expertise needed to apply high-performance design approaches more widely.

2

High-Performance DDC System Architecture

DDC Systems Today

Many in the industry think of advanced DDC systems as those developed during the 1980s and now in widespread use. But this decade is seeing enormous advances in the technologies employed in DDC systems. For a designer to consider high-performance controls as a part of an HVAC system design, the first step is to understand the changes in DDC systems over the last few years. From a hardware perspective, the more advanced systems are much simpler. They are easier to install, start up, calibrate, operate, and troubleshoot. Operators with basic skill levels and training can maintain modern DDC systems without outside help except for unusual problems. Users can also sign extended-warranty contracts that when procured competitively cost as little as $4 to $10 per DDC system point per year (in the 1980s, however, $50 to $100 per DDC point per year were not unusual costs for maintaining DDC systems). The costs of the systems themselves have fallen dramatically. In the 1980s designers often used $1000 per point to estimate the installed cost of DDC systems. When properly designed and competitively procured (not by low bid), high-performance DDC systems can today be fully installed for as little as $200 to $300 per point.

Elements of a High-Performance DDC System

Despite the wide variance in function among DDC systems today, an industry perception persists that all DDC systems offer more or less the same level of function, albeit in different fashions. This is a dangerous misconception for designers interested in implementing high-performance systems. Designers must understand that only a very few system configurations can optimally provide the high-performance control required for each specific project. Designers need to understand that there is never one best product or configuration available. Different DDC products and configurations may best match different projects depending on the size, functional requirements, operator interface requirements, and other factors. The following are the key elements of DDC systems that are rarely addressed by designers but are crucial to successful implementation of high-performance control. The most important elements depend on the specific requirements of the application. However, all these elements must be adequately addressed for a high-performance application to be successful.

A Fast and Efficient Communications Network

The distributed control concept of modern DDC systems has many benefits. But to be effective in a high-performance application, the distributed architecture must perform as a single integrated entity and not as a series of separate controllers. The communications network is the integrator. It provides features to make every system point or variable accessible for programs, trends, or displays in controllers and operators' terminals throughout the system. The communications network must have capabilities to ferret out the required information automatically, without requiring operator setup or involvement. How well a DDC system operates for the operator as an integrated entity is largely a function of the speed and effectiveness of the communications network. As the size of a distributed system grows, reliance on the communications network increases. A fast, fully automatic communications network is an absolute necessity for successful high-performance DDC system configurations.

Unfortunately, many manufacturers of DDC equipment have regarded communications as the least critical aspect of their design challenges. With the most crucial time-response requirements of typical HVAC controls being a few seconds or more, it has appeared unnecessary to manufacturers that the industry should be pursue the high-speed communications trunks such as those used in other computer applications. As a result, in high-performance applications, the DDC system network has often been the weakest link in the overall DDC system design. As discussed in Chap. 1 and later chapters, each controller operating terminal VAV box in a high-performance application may require the continuous exchange of the value or status of up to 50 system points and variables with other controllers on the network. A medium-sized building may have 500 terminal units. This means that 12,000 to 15,000 values or statuses must be continually exchanged over the network just to maintain terminal box operation. How the network is configured to obtain this throughput has much to do with how well the system will perform in a high-performance application—or whether it will perform in that application at all. Until recently, nearly all DDC systems employed standard RS232 communication trunks with maximum speeds of about 9600 baud (Bd) (9600 bits/s). Such network speeds are far too slow to provide integrated control in any but small network configurations. Today, manufacturers developing DDC systems for high-performance applications are employing networks that communicate at megabaud (millions of bits per second) speeds. These improvements are enormously increasing the network capabilities of DDC systems for high-performance applications. But the problems associated with networking requirements cannot be entirely solved by faster communications speeds.

Another critical feature of a DDC network is that it must be fully automatic. Some DDC systems require the programmer to list or tag the points that are required to be transmitted to other panels. Other systems require the program to denote any point that is used in the program but resident in another controller. These underdeveloped networks are not acceptable for high-performance controls. Imagine the effort required for the operator to oversee such a network when the networked points total in the tens of thousands!

Working with a DDC system that employs a truly automatic network, the operator need only enter a program, log in, or display any panel with valid point names. The energy management system (EMS) does everything necessary to develop the required network such that the value or status of points or variables resident in other panels is provided automatically to that program, trend log, or display as if they resided in the panel or console being programmed.

The automatic network is a feature that simplifies system operation. The DDC system determines the network required and through one of a variety of methods keeps each panel updated with information needed from other panels so that current data are available at the panel when needed. Automatic networking is essential for efficient communications, because the DDC system can organize a large network more efficiently than an operator can. Automatic networking is essential for high-performance DDC applications because it frees the operator from the tedious task of organizing and maintaining the network. Automatic networks simplify system operation in addition to providing more efficient communications among controllers.

Several different techniques are employed by automatic networks to update the value or status of each point that is required in other controllers. One method that has not been successful in large networks is to simply stop execution of the program and go to the network to retrieve the value or status of the point or variable located in another controller. This technique often causes an unacceptable program execution-speed reduction. Approaches better suited to high-performance applications employ buffers in each controller that are updated periodically with the value or status of points in other panels used by programs in this panel. One method of updating this buffer provides updates based strictly on a time interval. Points whose value or status is requested from other panels are updated to the buffer at specific time intervals (usually a few seconds). A second method of updating the buffer provides updates on "change of value." In this method, each digital point updates when it changes state, and each analog point updates when its value changes by an amount greater than a limit established in the point's definition.

A time-interval update approach works best in DDC applications because network traffic is much easier to control. Updating the value or status of each networked point every few seconds is well within the capacity of faster networks. In the change-of-value update scheme, chaotic network behavior can occur if conditions start changing rapidly or sensors become unstable. Furthermore, the change-of-value scheme does not work well for control strategies in which control actions are based on recent trends. Trends in temperature and other features show up as a series of steps in a change-of-value scheme that is difficult to use in a control algorithm. However, if changes in value or status must be relayed in faster than a few seconds (as is the case of some fire or security applications), the time-interval approach may not meet system requirements. Thus, a time-interval approach coupled with features that ensure more immediate response of critical point changes is the direction that the most effective DDC networks are taking.

A Powerful and Flexible
Operators' Control Language

When DDC systems first emerged, it was widely held by the system manufacturers that "canned" programs with simple input parameters were all a building operator needed (or, more to the point, all a building operator could handle) to implement and support DDC strategies. Users' complaints and demands for more functional systems have gradually destroyed the canned-program myth. Today, the need for powerful and versatile *operator's control language* (OCL) in DDC systems is universally recognized, although there are several views about the form that applications programs should take to be most functional.

Each manufacturer has an individual philosophy regarding the multitude of individual items that constitute a control language. Because of this, the designer must determine how competing systems differ with regard to their OCL and what tangible advantages one may have over another for the specific application. To develop an answer to this question, the designer should carefully guide an analysis of the OCL characteristics for each competing system. An OCL guide that covers the features required for high-performance DDC applications is included in the Appendix. The crucial issues that must be carefully considered in high-performance applications are as follows:

1. The OCL must offer control for each point through a single, comprehensive program.

The control program for each point must meet all the requirements of the application. This technique is called *output-oriented* programming, and it has been shown to be the most effective method to organize and support high-performance applications programs. Many DDC systems still utilize a combination of different program modules, or interlocks, or computed points, or optimization programs, etc. In these systems, the operator often has difficulty determining which module is controlling a point at any given time. Therefore, systems whose OCL is organized to provide control by software modules that are not output-oriented are difficult to operate and are not suitable for high-performance applications.

In output-oriented programs, every factor which establishes the operating characteristics or value of each point is located together in one program or section of the program. If the operator suspects a problem, it is a quick and simple matter to examine the portion of the program that operates the point in question to determine exactly how and why the point is being operated as it is.

2. The OCL must offer powerful and effective program editing features.

Entering or changing control programs in a high-performance DDC system must be as easy as editing a document with a modern word processing program. Some DDC program editors are slow and difficult to employ, making them unsuitable for high-performance DDC applications. Program editing must employ a full-screen editor that permits the operator to make required program changes with ease. Database save and reload features must be fast, offering entire system backups or reloads at under 0.5 s per system point. The editing features must include the ability to directly load OCL program code written off-line.

3. The OCL must provide a full range of mathematics and logic functions and permit flexible operator overrides for all system points and variables.

Every system point must have the capacity of being overridden by the operator at the console in a simple and uniform manner. A special character should denote the override condition each time the point is accessed or displayed. Some systems fail to provide this feature completely. If the outside air temperature sensor should fail, these systems do not permit the operator to easily override the point with a realistic value. Some systems do permit overrides of all points and variables, but do not provide a clear indication of the override condition each time the point is displayed. The override indication is essential in high-performance applications to serve as a reminder to return the point to automatic operation when the problem is corrected.

Adequate Memory and Effective Memory Management Techniques

Memory poses the same problems to DDC system manufacturers as it does to computer manufacturers in other industries—they have a tough time telling how much is enough. In DDC systems, point databases, applications programs, and trend logs all compete for available memory. The first releases of DDC products have been notorious for having severely limited memory. Newer versions of DDC products almost always include additional memory or memory-upgrade capacities that permit users to fully utilize valuable software features. However, available memory is sometimes allocated only to specific features, and there are often limits on the number or size of these features regardless of the memory increases made.

New memory management features in some DDC systems alleviate the problem of limited memory for specific functions by allocating reserve memory to specific areas as they fill the available memory. To ensure that the memory capacity of a DDC system is adequate for a specific application, the designer must know far more than the total number of bytes of memory available in each panel. The designer must understand how the memory is allocated to the various functions and what memory allocation techniques are available to obtain additional memory for specific functions. The designer must also understand the limitations on the number or size of specific functions that are not directly related to memory. Addressing schemes or other factors may limit features below those levels that could be supported by memory. All manufacturers have such information available in one form or another. But many dealers or branch offices provide rules of thumb that may not apply to high-performance DDC applications. The designer of a high-performance application must review the memory allocation and sizing rules for each product being considered to see how well the product will meet the requirements of the project. It is in the area of the amount of capacity (maximum numbers of programs, trend, etc.) that some of the largest differences among the various products become apparent. Such differences will not receive adequate attention unless the designer leads the effort.

Fast, Comprehensive Graphic Capabilities

Graphic displays of real-time point data and historical-trend data are essential to maintain the operator's ability to understand and manage the growing supply of information available from high-performance DDC system configurations. DDC system point counts are rising to the thousands and tens of thousands for high-performance applications in medium- and large-sized buildings. The operator must have rapid access to point and program information in order to manage the system's operation and use it effectively as a maintenance tool. A graphic menu approach is an excellent means for an operator to penetrate a large DDC system and isolate problems or concerns. But to be effective, the graphics must be fast—providing the complete graphic display in well under 1 s from the time it is selected. This is well beyond the capabilities of many DDC systems, but manufacturers are working to improve the speed of their graphics interfaces. Graphics capabilities must also be comprehensive, permitting operators with suitable security levels to access real-time or historical-trend information, programs, or database elements at any level of a single graphic menu system.

Reliable Operation

A few years ago, stand-alone panels had a reputation for being sensitive to power line problems. Elevators and emergency generators were but two of myriad reasons (a better word might be excuses) given for unexplained DDC system crashes and other failures. In recent years, the reliability of stand-alone panels has improved remarkably. Today, it is not unusual to find large systems operating for several years without a failure serious enough to require reloading a single panel. The issue of reliability usually emerges when manufacturers introduce new products. Designers must ensure that the system selected for a high-performance control application will offer reliable as well as functional operation.

DDC System Configuration

Trends in DDC system architecture have closely paralleled trends in industrial applications for computer-based control technologies. DDC systems continue to become more powerful and further distributed. Most modern DDC system architectures employ three types of components: the operator-interface console, input/output devices, and direct digital controllers which include *stand-alone panels* (SAPs) and *unit controllers* (UCs). The controllers (SAPs and UCs) are the heart of a DDC system. To be certain a system will perform as expected in a particular application, the direct digital controllers require a high level of understanding by designers.

Stand-Alone Panels

Stand-alone panels make up the nucleus of most current DDC systems. SAPs competitively purchased usually cost about $3000 uninstalled. How the SAPs perform is the single most important factor determining the overall performance of these DDC systems. Figures 2-1, 2-2, 2-3, and 2-4 show most of the DDC system architecture approaches in use or being developed today. In Figs. 2-1 and 2-2, input and output DDC points are interfaced to both SAPs and unit controllers. The difference between the two is that operator-interface devices are connected directly to an RS232 port of the SAP in Fig. 2-1 and to the high-speed communications trunk in the Fig. 2-2 architecture. The direct trunk connection usually means that data transfer to operator interfaces will be much faster. The architecture in Fig. 2-1 is more typical of DDC system architecture. However, in large DDC configurations, the direct trunk connection may significantly improve the speed for operator-interface features such as graphic displays. Direct trunk connection also decreases the time required for system backups or reloads.

The architecture in Figs. 2-1 and 2-2 has evolved from the development of functional stand-alone panels in the 1980s. As DDC systems have been expanded to include terminal unit control, a new type of controller was developed and connected to the SAP on a subnetwork. In this configuration, SAPs are employed in areas of high point densities such as mechanical rooms because these panels have quite large input/output point capacities. A SAP typically has the capacity to connect directly to about 50 total input and output points, and most also have the capacity for expander boards or slave I/O multiplex boards that can increase the point capacity almost indefinitely. The limiting factor in expansion is usually the amount of panel memory available to support applications programs, trend logs, displays, etc., for the points connected. In nearly all DDC systems, the SAPs are also required to tie the system together. As shown in Figs. 2-1 and 2-2, a SAP also controls the subnetwork for unit controllers. A SAP can generally support from 50 to 200 UCs on one or more UC trunks. Each SAP usually can also support one or two operator-interface devices. When procured competitively, stand-alone panels typically cost the end-user about $2000 to $5000 each (uninstalled) for a 50-point configuration.

More recent DDC system architectures are shown in Figs. 2-3 and 2-4. In these figures, the SAP has disappeared. In Fig. 2-3 it has been replaced by a *communications controller* (CC), and in Fig. 2-4 it has become an equal with the unit controllers in a full peer-to-peer network in which all controllers communicate directly with one another.

Figures 2-3 and 2-4 illustrate an important process in the evolution of DDC systems. Manufacturers have begun to recognize that the subnetwork architecture of Figs. 2-1 and 2-2 is an impediment to efficiently providing the high data-exchange levels required for high-performance applications. The architecture in Fig. 2-3 represents the manufacturer's recognition that fast, efficient communication is crucial to DDC system operation. The architecture in Fig. 2-4 is evolving as manufacturers determine economical methods of providing the required communication levels such that the communication controller becomes an integral part of each individual controller.

In Fig. 2-3 the communications controller has no input or output points directly connected to it. Designers should keep in mind that for

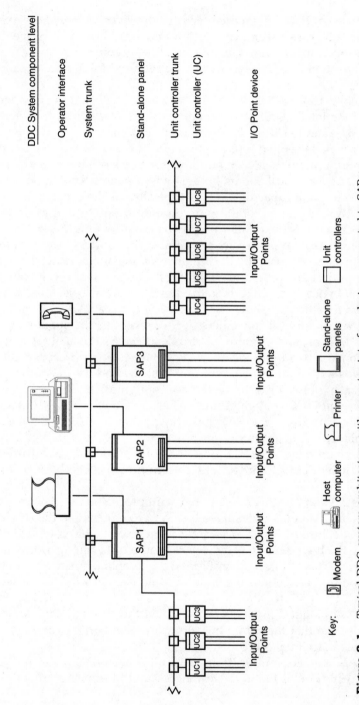

Figure 2-1. Typical DDC system architecture with operator-interface devices connected to SAPs:

DDC System component level

Operator interface

System trunk

Stand-alone panel

Unit controller trunk
Unit controller (UC)

I/O Point device

Key:
Modem   Host computer   Printer   Stand-alone panels   Unit controllers

SAP1   Input/Output Points

SAP2   Input/Output Points

SAP3   Input/Output Points

UC1   UC2   UC3   Input/Output Points

UC4   UC5   UC6   UC7   UC8   Input/Output Points

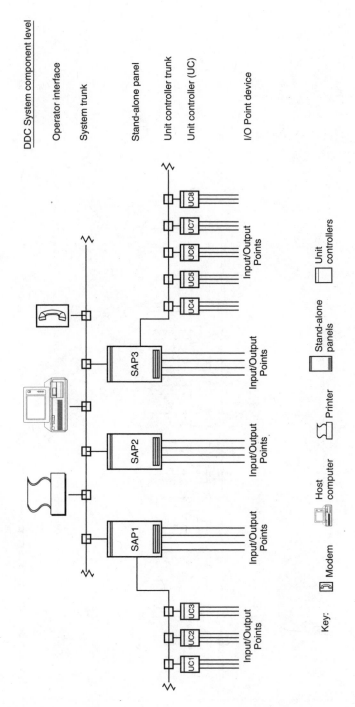

Figure 2-2. Typical DDC system architecture with operator-interface devices connected to the communications trunk.

DDC System component level

Operator interface

System trunk

Stand-alone panel

Unit controller trunk
Unit controller (UC)

I/O Point device

UC8
UC7
UC6
UC5
UC4

Input/Output
Points

SAP3

Input/Output
Points

SAP2

Input/Output
Points

SAP1

Input/Output
Points

UC3
UC2
UC1

Input/Output
Points

Key: Modem   Host
computer   Printer   Stand-alone
panels   Unit
controllers

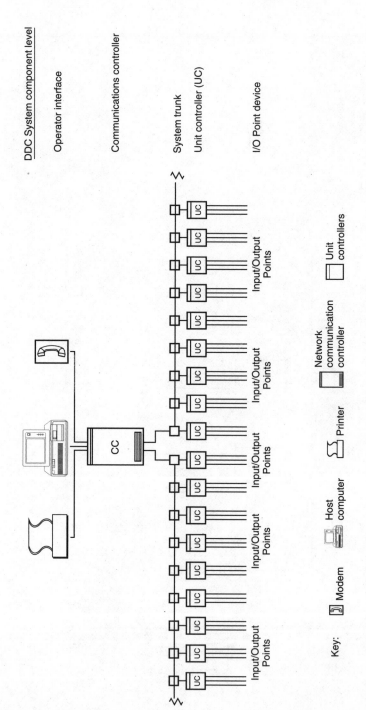

Figure 2-3. DDC system architecture employing communications controller.

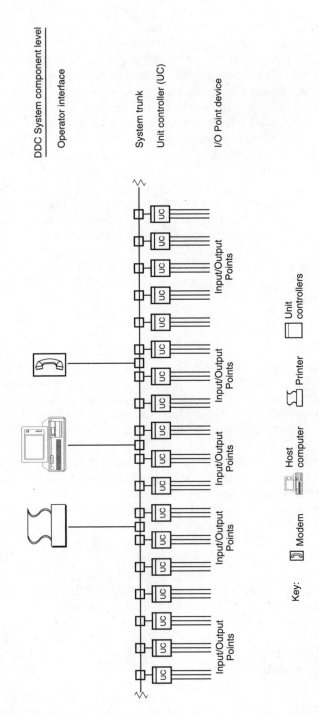

Figure 2-4. Emerging peer-to-peer DDC system configuration.

31

most DDC systems, this device acts very much like the SAP. Manufacturers that offer this architecture can usually provide either a SAP or a CC at this level, and the rules for number of unit controllers and operator I/O are generally the same for each. That is because the CC is almost always the exact SAP circuitry except it is assembled without any input or output device capabilities. The architecture in Fig. 2-4 illustrates the effects of the ongoing development of computer integration technologies. Using advanced integration technologies and emerging industry standards, DDC system manufacturers are now beginning to find that greater function can be obtained by installing the CC functions on each controller, and it is becoming economical to do so as well. This architecture development, which is now widely envisioned but not yet widely available, is already having an enormous impact on the direction of the DDC industry. There are a number of advantages to the DDC architecture shown in Fig. 2-4 that are sufficient to believe that this may be the architecture of the future for DDC systems. The major advantages of the DDC architecture in Fig. 2-3 are as follows:

Consistency

How much consistency is required throughout a DDC system with regard to point operation is a question the industry has been reticent to ask because the answer is so obvious and not what the industry would like to hear. How many different types of point interaction will an operator accept? The answer is not a half-dozen, four, or three. The answer is *none!* Every point throughout a DDC system—especially points of the same type such as a digital input—must be accessed exactly the same by the operator and subject to exactly the same rules when defining the point's database, calibrating, commanding, programming, overriding, or interrogating the point. Even small differences cause big problems. The two-tier architecture of Figs. 2-1 and 2-2 makes consistency a very serious problem. Even though manufacturers may offer "similar" means for dealing with points at the different levels, there is no DDC system for which the rules and procedures are *exactly* the same for points that are connected to SAPs and UCs. However, with all points connected to devices at a single level, the problem is much simplified. As long as the UCs all behave the same, the points connected to them can be made to operate relatively easily according to identical rules and conditions.

Economy

Typically unit controllers offer about one-quarter of the number of point connections and memory capacity of a SAP and cost about one-tenth as much. Newer unit controllers have the same level of functional capacity because they are usually derivatives of the SAP product. The reason for the lower costs is not exactly clear, but what it means is that UC-based systems are usually cheaper than SAP-based systems.

Congruence

Typical HVAC systems and components are usually better suited to be connected to controllers with lower point counts than SAPs. Controllers with smaller point capacities are well suited to connect a single air handler or chiller. Dedicating a unit controller to each HVAC component is becoming an increasingly popular DDC configuring concept. So long as an effective communication network connects controllers together, there appears to be economy in employing smaller controllers with no loss in functional capabilities.

SAP Structure

There are several approaches to structuring SAPs. One approach adopted by manufacturers employs a fully configured SAP on a single printed-circuit board. Figure 2-5 illustrates how a single-board SAP is typically configured. The idea is that a DDC system comprised of fewer parts is easier and more economical for the manufacturer to make, the distributor to stock, the contractor to install, and the user to maintain. Indeed, experience shows that single-board panel configurations often provide the lowest first cost. To make certain the panel serves a variety of applications, most single-board panels permit DDC connection points to be configured with great flexibility. Each point can typically be analog or digital, but output points and input points are not interchangeable. The downside to the single-board approach is that any component failure requires complete board replacement. Although plug-on terminations and fast reload procedures make panel replacement nearly as fast as replacing a single component, it can be expensive. As long as failures are few, and manufacturers or third-party firms provide economical board repair, the single-board approach will be successful.

A second approach is to build a panel from modules. Separate mod-

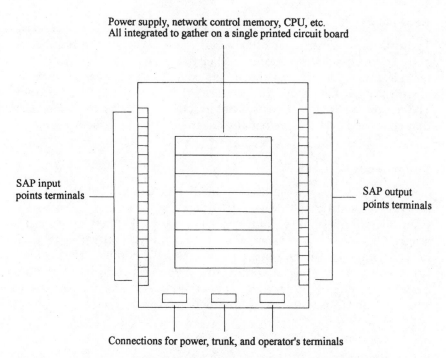

Power supply, network control memory, CPU, etc.
All integrated to gather on a single printed circuit board

SAP input
points terminals

SAP output
points terminals

Connections for power, trunk, and operator's terminals

Figure 2-5.   Single-board stand-alone panel (SAP) configuration.

ules are inserted into a mother board to configure each input or output, and only those modules that are required for the actual I/O points in each panel are installed. Power supplies, communication controllers, and other functions are also modularized by some manufacturers. A diagram of a modular SAP is shown in Fig. 2-6. Some advantages exist for a modular SAP because it can be configured to meet the exact I/O combination required. However, because the sum of the individual modules tends to be more costly than the single-board panel, the real advantage of the modular approach is the enhanced ability to isolate and repair failures quickly and inexpensively. How much of an advantage this can become depends on the costs of troubleshooting and replacing a single module compared to replacing the whole panel.

A third approach to SAP hardware is to make a compromise between the panel on a board and the modular panel. Manufacturers adopting this approach wish to use advantages from both the others. The semimodular panel has fewer modules than shown in Fig. 2-6 and most of the quick-change panel features in Fig. 2-5. Each

Expansion memory in plug-in module

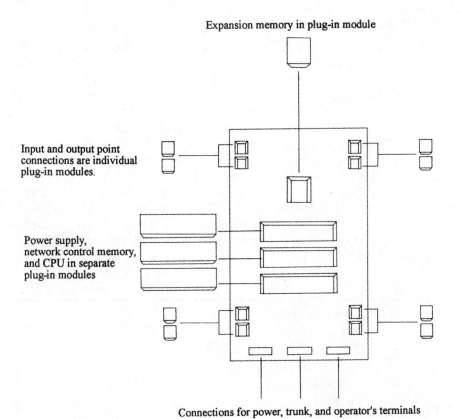

Input and output point connections are individual plug-in modules.

Power supply, network control memory, and CPU in separate plug-in modules

Connections for power, trunk, and operator's terminals

Figure 2-6. Modular stand-alone panel configuration.

approach has both advantages and disadvantages compared to the others when various features are considered. The differences among all the approaches are certainly not significant. I have never found any compelling reason to specify or favor one of these types of panels over another. The most effective means of determining the right panel for a given application is to require a long-term extended-warranty proposal along with the DDC system price. The system cost added to the present value of the extended warranty provides the best picture of the true system cost over time. Whether or not the user intends to sign an extended-warranty agreement, a warranty proposal from the vendor in a competitive environment is probably the best estimate of the true cost of maintaining the system. In the final analysis, life-cycle cost is the primary reason for selecting one hardware configuration over another.

SAP Function

SAPs typically contain all the capabilities needed to provide DDC control, including coordinating communications among panels, executing control programs, PID loop control, storing trend log values and status, etc. Operator-interface devices can all be disconnected without affecting automatic control in true stand-alone panels because all real-time functions are contained at the panel level.

Although manufacturers have adopted a variety of philosophies regarding operating features, a number of features have been universally recognized and incorporated in SAP designs. Figure 2-7 shows the checklist of SAP features and attributes that designers should review closely to determine if a particular DDC product will be successful in a high-performance configuration. These features are crucial to achieving success in each high-performance DDC system application.

Communications network
1. High speed
2. Auto build
3. Efficient updating

Operator's control language
1. Output-oriented
2. Powerful editor
3. Full range of mathematics, logic, and control functions

Memory and memory management
1. Sufficient total memory
2. Adequate memory management for allocation
3. Adequate numbers of every required function

Operator interface
1. Adequate speed and flexibility
2. Suitable for all who access system

Reliability

Consistency
1. Uniform access of all points
2. Economy
3. Congruence of architecture to application

Figure 2-7. Checklist of DDC features required for high-performance applications.

Unit Controllers

The relatively recent development of unit controllers that emulate SAP functions offers new opportunities to HVAC designers and system operators. Unit controllers typically offer limited point capacity (8 to 16 inputs and 8 to 16 outputs) and usually have no expansion capacity. Uninstalled, a unit controller typically sells for $100 to $300. Because unit controllers were originally designed with the idea of controlling small actuators and devices, some come with only digital outputs, in which case two physical outputs are employed to control electric floating-point-type modulating actuators. Unit controllers are almost always single-board products. Examples of how a unit controller may be connected for a variety of HVAC control applications are shown in Figs. 2-8, 2-9, and 2-10.

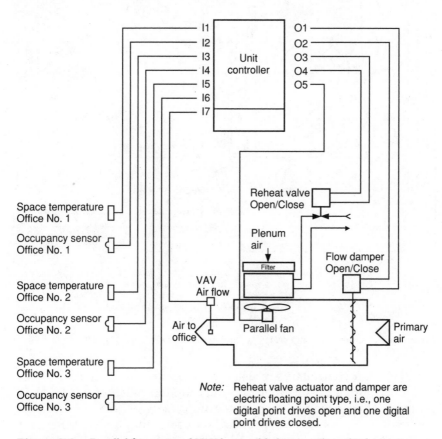

Figure 2-8. Parallel fan powered VAV box and lighting with multiple zones.

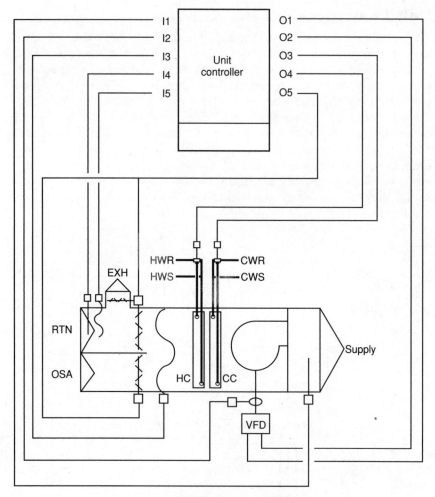

Note: Mixed air damper actuator, cooling
valve actuator, and heating valve
actuator are electric analog type,
i.e., analog voltage output determines
position of actuator.

Figure 2-9. Unit controller in HVAC fan system control.

Today, unit controllers are usually configured on a separate trunk
that is supported by a SAP or CC. The communication of this trunk is
generally slower than the communications network that connects the
SAPs and does not employ peer-to-peer communications. Depending
on the manufacturer, thirty to several hundred unit controllers can, in

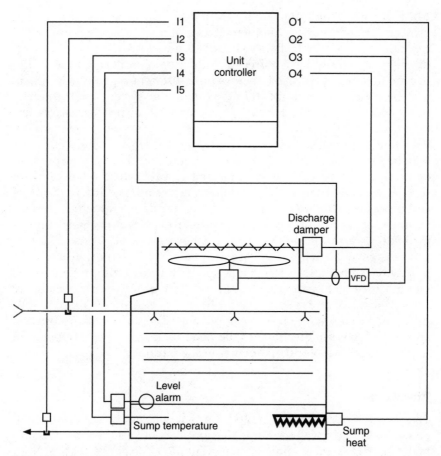

Figure 2-10. Unit controller in cooling tower control.

theory, be attached to a SAP or CC. Figures 2-1, 2-2, and 2-3 show how unit controllers are configured in DDC systems today.

The UCs available today have several problems that can cause considerable headaches for DDC system operators who have grown accustomed to the power, flexibility, and simplicity of SAPs. Unit controllers can cause these problems:

Consistency

The issue of consistency of point operation was discussed earlier as a potential benefit of a single-tier DDC architecture, as shown in Fig.

2-4. However, single-tier architecture does not necessarily solve the problem of consistency. Manufacturers often employ what are called *application-specific* controllers to handle the variety of different control functions that may be required of individual controllers in a single-tier architecture. Figure 2-11 shows how different types of controllers may be connected and operate on a single-tier system. Such application-specific controllers may not use consistent processes to command, calibrate, or program points. They may not even be made by the same manufacturer. For example, the chiller or boiler controller may be developed by a different division of the company or perhaps by an outside manufacturer. Replacing or calibrating a temperature sensor associated with one controller may require a different device and procedure than replacing one that performs the same function for another. The programming procedures for calculating setpoints or manually overriding points may be different as well. These different procedures, even when they are only slightly different, add a complexity to the operator's job that is not at all well accounted for by the manufacturers of such systems and is, for the most part, completely unnecessary. Manufacturers who have developed techniques that make their stand-alone panels simple and straightforward to operate appear to have underestimated the need for consistency of operation to meet these goals for their smaller unit controllers.

Function

The problem of consistency noted above would not be so serious if all one wished to do was to mimic the simple control functions of pneumatic control devices. But replacing simple control approaches with no added function is not a compelling reason to upgrade to DDCs. The reason why most designers and users desire to move up to full DDC systems is to provide more functional control of all HVAC components in order to improve the comfort and efficiency of the mechanical system. Figure 2-12 illustrates a common control problem. A group of small offices is supplied by a single variable-air-volume (VAV) box. Traditional controls would place a single space thermostat in one office to operate the box volume damper and reheat coil with the hope that all offices have consistent heating and cooling loads and once balanced will maintain a level of comfort similar to that of the controlled office. Building engineers know all too well that this type of control is unsatisfactory. Sooner or later an occupant in one of the rooms without the thermostat will complain about comfort. The building engineer has no satisfactory means to solve such problems. Can

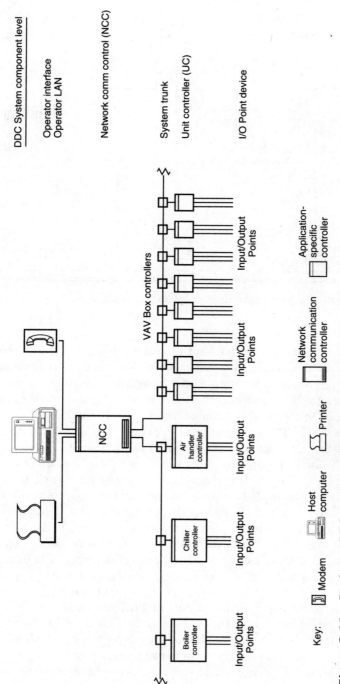

DDC System component level

Operator interface
Operator LAN

Network comm control (NCC)

System trunk
Unit controller (UC)

I/O Point device

VAV Box controllers

Input/Output Points

Input/Output Points

NCC

Air handler controller

Input/Output Points

Chiller controller

Input/Output Points

Boiler controller

Input/Output Points

Key:

📞 Modem

🖥 Host computer

🖨 Printer

Network communication controller

Application-specific controller

Figure 2-11. Single-tier DDC system architecture employing application-specific controllers.

41

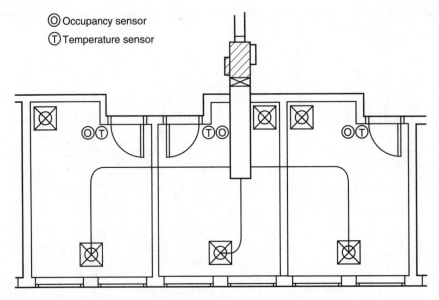

Figure 2-12. Single-terminal units supplying three subzones.

the thermostat be moved without offending that occupant? Would a rebalance of the outlets to send more or less flow to the offending space be anything more than a short-term fix? These are situations that designers have not had the tools to resolve, and they result in operating problems that are almost never entirely solved.

A functional DDC system offers opportunities to provide a more comfortable, higher-quality environment in such an HVAC configuration and at the same time to improve the energy efficiency of the system. The DDC points shown in Fig. 2-12 show how such improvements can be accomplished. DDC system space-temperature sensors are very low-cost items, making it feasible to install a temperature sensor in every office. To control lighting and the HVAC terminal unit, each office might also employ a push-button or occupancy sensor to notify the DDC system that the space is occupied. If a push button is the occupancy device, occupancy may be assumed to be continuous for the day when activated during normal business hours or for an agreed fixed time if activated outside business hours. An occupancy sensor offers further control by providing a signal only when someone is in the room. A short delay in switching to unoccupied condition after the room is vacated permits stable operation.

In Fig. 2-12, the occupancy signal controls lighting for each office.

The occupancy signal is also used to operate the VAV terminal box that serves all three offices. If only one office is occupied, then the box operates to satisfy that office. When multiple offices are occupied, the box operates to satisfy the average of the occupied offices unless an office exceeds high or low space-temperature limits; in that instance, the box would act to bring the office temperature within limits, then return to averaging control. A typical control program to operate a VAV box in this manner with multiple space-temperature sensors is shown in Fig. 2-13.

With the occupancy signal from each room as an input to the DDC system, the box need have no set minimum airflow. If all rooms are unoccupied, the box may shut off entirely so long as all offices are within suitable unoccupied temperature limits. When one or more spaces are occupied, the minimum airflow is set by the DDC system to provide the required outside air to the zone—based on an operator-entered value for the number of occupants in the space when occupied. The DDC system can calculate the minimum airflow required from the percentage of outside air being supplied to the box at the time.

Such an operating sequence, together with additional features such as night purge cycles, can lead to comfort and air-quality levels for building occupants that are vastly superior to those provided by traditional pneumatic terminal controls. The marginal cost for the DDC system to implement better terminal control is increasingly compensated for by energy and operating-cost savings. A problem facing designers is that very few of the terminal-control products on the market can effectively accomplish the sequence described above. Most UC products require extraordinary database or program manipulation in order to do so, making such sequences difficult to implement and difficult for an operator to support.

Flexibility

When manufacturers began designing UC products, it appeared that the primary goal was to replace existing pneumatic controllers with a DDC product whose only advantage was remote monitoring. This unambitious goal has resulted in the industry being deluged with unit controllers whose application software is preprogrammed with simple control sequences in *read-only memory* (ROM). Many of the products available today lack both the functional capacity to solve the design problem outlined above and the flexibility for the operator to make even simple changes or adjustments in their sequence of operation once they have been installed. Any change the operator wishes to

```
DOEVERY 1 M
    IF LIGHT_1 = ON THEN BEGIN
        IF ST1 BETWEEN HTGSPA CLGSPA THEN BEGIN
            A2 = ST1
            B2 = 1
        END
        ELSE BEGIN
            A2 = ST1 * 2
            B2 = 2
        END
    ELSE BEGIN
        A2 = 0
        B2 = 0
    END

    IF LIGHT_2 = ON THEN BEGIN
        IF ST2 BETWEEN HTGSPA CLGSPA THEN BEGIN
            A2 = A2 + ST2
            B2 = B2 + 1
        END
        ELSE BEGIN
            A2 = A2 + ST2 * 2
            B2 = B2 + 2
        END
    END

    IF LIGHT_3 = ON THEN BEGIN
        IF ST3 BETWEEN HTGSPA CLGSPA THEN BEGIN
            A2 = A2 + ST3
            B2 = B2 + 1
        END
        ELSE BEGIN
            A2 = A2 + ST3 * 2
            B2 = B2 + 2
        END
    END

    IF B2 > 0 THEN
        BOX_SPACE_TEMP = A2 / B2
    ELSE
        BOX_SPACE_TEMP = AVG(ST1 , ST2 , ST3)
ENDDO
```

LIGHT_1, LIGHT_2 and LIGHT_3 are the status of the lighting output relays for each of the box's subzones.

HTGSPA, and CLGSPA are the current heating and cooling setpoints.

ST1, ST2, and ST3 are the values of the space temp sensors in each subzone.

BOX_SPACE_TEMP is the weighted average space temperature.

A2, and B2 are local program variables.

Figure 2-13. Program for averaging multiple space-temperature sensors.

make that is not a preprogrammed option cannot practically be accomplished at all in many unit controllers.

Some designers believe that ROM-based backup programs in unit controllers are necessary in case of network failure. Although once this was sound advice, ROM application programs at any level of a DDC system are no longer necessary. New memory technologies such as EPROM and "flash" memory programs allow programs to remain intact indefinitely without power. ROM-based programs might be used to ensure that certain interlocks to protect compressors or fans are not removed, or to prevent damage from short cycling. However, these strategies are based on the concept that the system must be protected from the operator, which has been discredited as a means to attain high-performance operation of virtually any system.

To offset limitations from UCs that operate ROM-based programs, some manufacturers have built in the ability to "unbundle" UC points and operate them as if they were points connected to the SAP to which the UC is connected. There are two problems with the unbundling approach that make it unworkable for typical terminal-control applications. In most systems unbundling is a cumbersome process because each point that is to be unbundled must be redefined in the stand-alone panel. Still, in some instances the unbundled points cannot be operated the same as a panel points. Also, unbundled UC points add to the point count of the SAP, which can severely restrict the number of UCs that a single SAP can serve.

Opportunities Ahead

As more functional and versatile UCs are developed, the opportunities for improving the comfort and economy of HVAC systems will open to entirely new areas. The recent ASHRAE standard for ventilation has been criticized because it does little to ensure that ventilation air is actually delivered to each of the building's occupants. Using methods similar to the earlier unit control example, DDC systems can ensure on a zone-by-zone basis that adequate outside air is supplied to every occupant and at the same time can improve the comfort and energy efficiency of each building.

Meanwhile, the development of UCs and associated technologies is a current industry revolution that will eventually replace SAP-based systems with UC-based systems with UCs that act much as SAPs do today, but are less costly on a per-point basis and can be connected to a fast, efficient network in much larger numbers.

The prospect for expanding building DDC systems to terminal control is indeed very promising. But to be certain the promises regarding opportunities for full DDC are kept, the industry must encourage DDC system manufacturers to improve the function and flexibility of their UC products and to develop operational features more consistent with the SAPs which they will soon replace. Once such powerful products are available, designers can use the added power to improve building comfort, air quality, and cost-efficiency up to the levels that owners and occupants now demand.

Points to Remember

The basic architecture of any DDC system is established by the manufacturer's designers when the product is developed, and no two DDC systems employ identical system architectures. The DDC architecture can usually be adjusted somewhat to meet specific project needs by the applications engineers, but the elements of the basic architecture employed in a DDC system are very telling about how it will perform in a high-performance application. Designers need to be aware of the differences among the various systems proposed for a specific application. Designers cannot rely on branch office or dealer personnel to see that each DDC system proposed for an application is optimally configured for a high-performance application. Designers must develop procurement procedures to be certain that the DDC system which most cost-effectively meets the project's needs (not necessarily the lowest-cost system) will be selected.

To develop the background necessary for these duties, there is no substitute for hands-on experience. Designers cannot claim a good understanding of any DDC system if they do not have hands-on experience with each product. There are a number of methods by which such experience can be developed, but by far the best is working with knowledgeable building operators through the start-up process, while problems are being solved or improved strategies are being implemented. Operators are usually most receptive to the idea of working with the designer during start-up. Designers find that gaining experience about DDC systems in this manner is far more useful than discussing the product with the manufacturer or distributor.

3

DDC System Input/Output Devices

The significance of DDC system input and output devices has increased enormously as the development of terminal controllers has encouraged the use of full DDC system configurations. DDC systems that control only fans, chillers, and central plant equipment require as few as 100 to 200 input and output points or less to effectively control a medium- to large-sized building. Full DDC configurations that incorporate lighting and temperature control for every space require many thousands of points. This order-of-magnitude increase in points has changed the way designers must configure DDC systems to be cost-effective and to perform successfully. The importance of the input/output device cost has changed from a minor item to the leading factor in determining how cost-effective the DDC system will be. A current rule of thumb is that the installed cost of a full DDC system must be kept to under $200 per point to have a chance of being cost-effective when compared to other approaches from a purely first-cost standpoint. This objective can be achieved only if the system designer pays close attention to the selection of input and output devices.

The strong emphasis on input/output devices is required because in full DDC system configurations, these devices represent a large portion of the controls cost. Even more important, however, is that these devices are what link the DDC system to the building. The designer's task is to specify components that provide desired sensing or control characteristics reliably at the lowest possible cost. To do this, the designer must first determine what characteristics are realistic.

In selecting input/output devices, designers should remind themselves that if their full DDC systems are configured such that they cannot be economically justified over pneumatic alternates, then their clients and the building occupants are the big losers. Consider for a moment the comparative precision and stability of pneumatic thermostats and humidistats to modern electronic devices. If DDC systems cannot be configured that meet budgetary and cost-benefit requirements, the lower-cost alternative is usually pneumatic controls. But the performance capabilities of pneumatic devices are becoming inadequate to meet the growing needs and desires of building occupants in today's building environment.

This understanding of the role and importance of cost considerations is required for a designer to make reasonable choices for terminal input/output devices. Terminal output devices such as damper actuators and reheat valves have often been considered to be perfect applications for pneumatic actuators. However, when the applications are scrutinized more closely, pneumatic devices are often not a good choice at all for typical HVAC control applications. Furthermore, pneumatic devices do not interface well to electronic control systems. As a result, a number of new economical actuation devices have been growing in popularity over the last few years.

Temperature Sensors

The effect of temperature on the electric resistance of materials was discovered more than a hundred years ago, and *resistance temperature detectors* (RTDs) have been in use for well over fifty years. The earliest such RTDs employed platinum which is still used as a standard because of its extremely wide temperature range, stability, and resistance to contamination. More recently a less expensive device has been developed which is actually a type of RTD, but is more commonly known as a *thermistor*. Thermistors are generally constructed of semiconductor resistance materials, and most have a negative *temperature coefficient* (TC) whereas the metals used in RTDs have a positive TC. Thermistors are far more sensitive than the metal-based RTDs, having TCs that may be as large as several percent per degree Celsius. However, this sensitivity requirement is usually well beyond the requirements of usefulness in

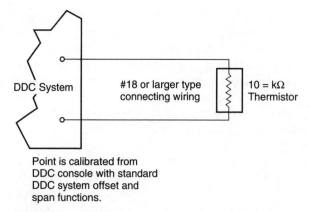

Point is calibrated from
DDC console with standard
DDC system offset and
span functions.

Figure 3-1. Typical thermistor temperature sensor connection.

typical HVAC control applications. Thermistors are far less linear than metal RTDs, and the stability of thermistors is limited at a lower temperature ceiling, usually about 150°C. Some long-term drift may develop at these ceiling temperatures. But linearity is not an issue for DDC systems because they almost universally have the capacity to employ lookup tables that "linearize" any type of nonlinear input. Furthermore, most HVAC temperature-sensing applications are well below the stability ceilings of thermistors, so by virtue of their very low cost, thermistors are well suited to HVAC applications.

A major benefit for thermistors over metal RTDs for DDC system applications is the fact that thermistors typically are much higher resistances, a 10,000-Ω thermistor being very common. This means that the thermistor can be connected directly to the DDC system without any additional device(s). Any effects of temperature change in the connecting wire resistance will be insignificant. This is not true of metal RTDs whose nominal resistance is often 100 Ω. To operate accurately, they require some circuitry to be located very close to the sensor or additional wiring and a compensating circuit at the controller. This circuitry (usually a bridge circuit) adds to the cost and adds more components that affect the overall precision and reliability of the device. Additionally, the intermediate circuitry of many RTDs does not permit calibration adjustments to be made easily in the DDC system. Calibration of the RTD usually has to be done by adjusting one or more trim pots on the device itself, further increasing their complexity. Figure 3-1 shows diagrammatically how the use of thermistors offers simpler DDC configuration and operation.

Precision RTD sensors of the type shown in Fig. 3-2 are still commonly specified for space-temperature sensors in DDC applications. Such devices can cost up to $100 or more apiece whereas thermistor-based sensors typically cost under $15. The reason why RTD sensors are preferred by designers for space-temperature sensing is not at all clear. In space-temperature-sensing applications, the limiting factor in accuracy is almost never the sensing device. The end-to-end accuracy of space-temperature sensing for DDC systems involves a number of considerations, the most important of which is usually the space itself. Consider that typical occupied spaces in buildings are supplied with air that is 20°F warmer or cooler than the space. As this air is circulated to mix with the space air in order to add or reject heat, temperature gradients within spaces are often significant. If several independent instruments are placed within a few inches of a space-temperature sensor, their readings will usually differ among themselves by 0.2 to 0.5°F. Thermistor-based space-temperature sensors cost only a fraction of the cost of precision RTD sensors yet provide stable operation and precision that are many times better than the 0.2 to 0.5°F precision that can be realistically achieved in space-temperature-sensing applications. There are not often valid reasons for disallowing the use of thermistors in typical HVAC applications except in high-temperature-sensing applications. Thermistors offer fast response, very high sensitivity, excellent accuracy and repeatability, and very low cost.

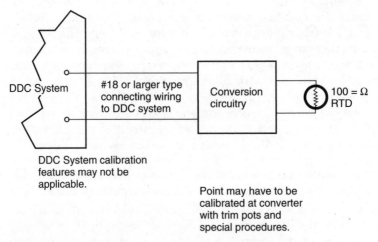

Figure 3-2. Typical metal RTD sensor connection.

Humidity Sensors

In the days of pneumatic controls, humidity control was rarely present at all in typical building-control applications. The applications for humidity control were largely limited to providing some low minimum level of building humidity in cold climates.

One of the reasons why engineers shied away from humidity control has been the level of technology employed to sense humidity. Most pneumatic humidity sensors were based on materials that expand or contract with changes in relative humidity. Nylon, cellulose, and human hair have been employed in various sensors. There were two problems with these sensors. First, strain devices are subject to drift and require frequent calibration at regular intervals. Second, strain-type humidity sensors are subject to stress relaxation, which meant that the device's ability to register changes in humidity deteriorated over time.

The current replacement of pneumatic controls with digital-based control systems for buildings has expanded the market for new humidity sensors that are based on electrical measurement of humidity. These sensors are being widely used because they have vastly improved long-term reliability and accuracy and are relatively inexpensive.

There are two common types of electricity-based relative-humidity sensors: the bulk polymer resistance sensor and the thin-film capacitance sensor. The bulk polymer type of sensor measures relative humidity by the resistance changes in a bulk polymer film. Humidity in the air causes ions to be released from the molecular film structure. As the relative humidity increases, the ionization increases which decreases the resistance of the film. The bulk polymer type of sensor is not generally susceptible to contaminants in the air, except chemical contaminants with properties similar to the polymer base employed in the sensor. These are seldom encountered in office environments. The bulk polymer sensor is not accurate at very low humidities, but generally provides very stable and accurate readings in the range of 30 to 90 percent relative humidity.

The thin-film capacitance sensor employs a humidity-sensitive material that absorbs water from the air to change its dielectric constant. The material is configured in a variety of fashions as an electronic capacitor that permits exposure of the material to air. The capacitance of the sensor is dependent on the relative humidity of the air. This capacitance-type sensor generally responds more quickly than the bulk polymer type, responding to changes in humidity in seconds instead of minutes. Capacitance sensors are accurate at low relative humidities, although they can become unstable at very high humidities. Generally thin-film capacitance sensors will provide accurate readings in relative-humidity ranges between 10 and 80 percent.

Both types of sensors are economical, and both types can be counted upon for accuracy and long-term stability many times better than could be achieved with pneumatic sensors. Typically, these types of space and duct humidity sensors are available in 3 and 5 percent accuracy. The 5 percent models cost from $100 to $200 and are entirely satisfactory for humidity control applications in most office and institutional buildings.

Pressure and Differential Pressure Sensors

Several techniques are employed to take pressure measurements in DDC system applications. The most common method is the use of capacitance. In this approach, two stainless-steel diaphragms are electrically isolated, and a circuit is provided to convert the measured capacitance to a voltage or current signal. One of the diaphragms is permitted to move toward or away from the other depending on the relative pressures in the chambers on each side of the diaphragm. The change in capacitance resulting from the diaphragm movement is converted to a voltage or current signal. A diagram of this type of sensor is shown in Fig. 3-3. Other methods of pressure sensing involve the use of semiconductor materials whose electric resistance changes with pressure or strain. Sensors that utilize materials whose resistance changes with pressure (called *piezoresistive*) are usually not as susceptible to mechanical shock or strong vibration.

Some of the most sensitive pressure sensors are constructed similar to the capacitance type except a silicon diaphragm is employed instead of a stainless-steel one. The resistance of the silicon diaphragm changes as it deflects. A diagram of this type of sensor is shown in Fig. 3-4. These devices provide accurate pressure measurements in full-scale ranges as small as 0.1 in for use in pitot-type air velocity sensors or for building pressure-control applications. The costs of any of these types of sensors are generally under $300.

Airflow Sensors

One of the most noticeable advantages of DDC terminal VAV control is the ability to provide air balancing of terminal boxes with the DDC

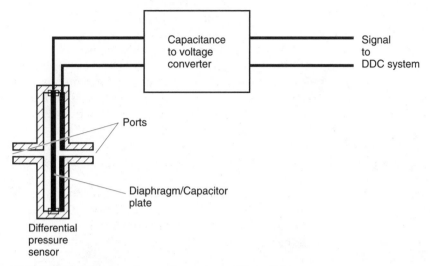

Figure 3-3. Diagram of capacitance-type differential pressure sensor.

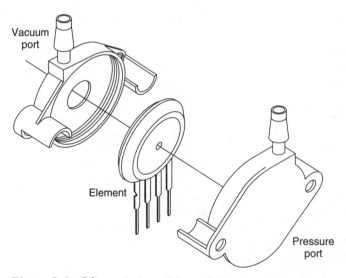

Figure 3-4. Silicon strain resistance-type pressure sensor.

system. DDC air balancing is much easier than traditional methods. The maximum and minimum airflows for each box are set by the DDC system, and each VAV box damper operates to provide a specific airflow within that range based on space conditions. Duct pressure compensation is automatic.

To achieve all these features requires an airflow sensor at each VAV box connected to the DDC system. At present there are two approaches employed by the HVAC industry to measure airflow at terminal boxes. Both actually measure air velocity which is multiplied by a constant (the constant includes factors for duct area and flow profile compensation) factor to obtain airflow. One approach employed to determine the air velocity at terminal units is to measure the velocity pressure of the air through the use of a pitot-tube arrangement and an air-differential pressure sensor. Some of these velocity pressure devices actually measure the average velocity pressure with a series of upstream orifices to help compensate for the irregular velocity profile that occurs in many ducts. The velocity pressure airflow measurement technique has one serious limitation. At low air velocities, the velocity pressure of air is very small. Much of the literature recommends that the velocity pressure technique be limited to air velocities above about 300 ft/min. Tests that I have conducted found reasonable accuracy somewhat below that figure under certain conditions, but the bottom limit of accurate flow measurement is still above the lowest velocities encountered at VAV terminal boxes.

Another problem with velocity pressure-type sensors is that the differential pressure/electric transducer required to convert the velocity pressure to an electric signal and the sensing equipment can be costly. Some DDC manufacturers include the differential pressure/electric transducer with their box controller which seems to dramatically reduce the transducer cost. Controllers incorporating the transducers are usually competitively priced with those not including transducers.

A second approach to measuring airflow is to determine the air velocity by measuring its capacity to provide cooling. Methods that employ this approach are sometimes referred to as the *hot-wire* or *heated-thermistor* techniques. There are a number of variations in the operation from manufacturer to manufacturer and depending on the exact devices incorporated in the design. Essentially, one device is heated, and some mechanism determines the cooling effect of air passing over it by measuring the actual temperature of the device (or the current required to maintain the device at a constant temperature). The cooling-effect approach has one big advantage over the velocity pressure technique: It can accurately measure air velocities as low as 50 ft/min. The most economical of the cooling-effect velocity sensors employs several thermistors. Exact operation varies, but a popular technique is to employ one thermistor to measure the ambient temperature of the airstream. The second thermistor (slightly downstream of the first) has a voltage applied to it, and the current required to maintain the voltage is measured. Because the resistance of thermistors varies with temperature, the temperature of the heated thermistor can be calculated. The combination of ambient temperature, heated-thermistor temperature, and energy to the heated thermistor can be used to calculate the air velocity in smooth flowing air. It may seem complicated, but these comparisons are usually made in a small IC chip. The cost of thermistor-based cooling-effect air velocity sensors is about $25 or less, and some automatically provide leads that can be connected to another input of the DDC system to provide the air temperature as well. Most do not provide a linear output. But most DDC systems offered today have the ability to linearize such signals with calibration tables included in their input point databases.

A disadvantage of the cooling-effect airflow measurement technique is that it provides a reading of airflow at only one point in the airstream. To read more points for average airflow (which is easily accomplished with the velocity pressure technique) requires an array of a number of individual sensors. Thus, a sensor designed to measure several points can become expensive. Our firm has conducted performance tests of airflow-measuring devices. The tests concluded that by following some basic rules for sensor location, a single-point velocity sensor can be employed to measure airflows in the duct sizes typically employed for VAV terminal units with adequate precision. In addition to sensor location, these tests found that how the velocity-to-flow factor was determined is the primary factor in achieving accurate measurements.

Because it generally offers the lowest-cost airflow measurement option, thermistor-based airflow measurement has a jump on other techniques for DDC terminal VAV box control. However, thermistor-based velocity sensors have pitfalls that designers should consider when selecting airflow devices. The low-cost thermistors employed in HVAC applications are generally stable and drift-free in applications up to about 300°F. Above this temperature many are subject to drift. Because the heated thermistor in airflow sensors usually operates close to the 300°F stability ceiling, some in the industry have expressed concern that periodic recalibration of the airflow sensors may be required. I have seen no conclusive evidence on the subject of drift for heated-thermistor sensors, but most of the installations I am familiar with are only several years old. Meanwhile, manufacturers are very short on facts that might prove drift will not be a problem. One solution to the drift problem is to employ sensors that use stable RTD sensors in hot-wire configurations. Another is to select higher-quality thermistors that are not subject to drift at elevated temperatures. Glass-encapsulated thermistors are generally very stable at higher temperatures. Recent entries into the air velocity sensor market do utilize glass-encapsulated thermistors and are priced around $25 per unit.

Electric Current Sensors

A device that is becoming very popular to indicate the status of fans, pumps, and other power equipment is the electric current sensor. There are two type of sensors. Both are current transformers that are affixed around one of the power wires to the device being monitored. The digital-type sensor has a solid-state switching device included that switches status at an adjustable current level. In many of these devices, the differential current is adjustable also. The analog-type current sensor has a converter that provides a voltage or current signal to the DDC system proportional to the current measured. These current devices provide far better reliability than the flow or pressure switch in fan and pump monitoring applications because they employ no moving parts. Because the current will fall significantly if a belt or coupling fails, these current devices provide a much more complete status signal than an auxiliary starter contact.

Of these two devices, the analog device is preferred in most DDC system applications because it provides so much more information than just the status of the unit. For the analog current sensor, a low-current alarm can be entered in the DDC system to provide a failure alarm. If the alarm is triggered, the operator can tell by looking at the current if the motor has failed or the belt is broken. Additionally, a high-current alarm can be implemented for the same device which will indicate if the system is in an overload condition. Current-sensing devices are somewhat more costly than the air and water differential pressure switches typically used for fan and pump status indication, but the installation cost is usually less, especially when the starters for a number of the fans and pumps are located in a single-motor control center.

Occupancy Sensors

Incorporating lighting control with HVAC control offers substantial improvements in energy efficiency and lighting system performance over stand-alone lighting control systems. To effectively sense occupants for lighting and HVAC control, the designer has several options. The simplest type of occupancy-sensing device is a button on the space-temperature sensor that the occupant pushes upon arrival in the zone. When the button is pushed, the DDC system turns on the lights and establishes an HVAC occupancy mode for the zone. Typically, the zone remains in the occupied mode for the remainder of the day if the button has been pushed during normal working hours or for a predetermined time if it is activated during off hours. The most effective strategy permits the button to act as a toggle; successive presses of the button toggle the zone back and forth between occupied and unoccupied states. Many manufacturers now incorporate such a button in versions of their temperature sensor. In some DDC systems the push button is multiplexed with the temperature sensor input and does not require additional input point or wiring. The result is that a push-button occupancy device can be very inexpensive to provide—adding almost no cost to the system.

More effective means of zone occupancy control can be implemented with the use of automatic occupant-sensing devices. These devices sense movement with either infrared or ultrasonic sensing mechanisms. Infrared occupancy sensors pick up movement by sensing heat from people as they move. Ultrasonic occupancy sensors send out a very much lower level of ultrasonic energy and react to changes in reception patterns that is due to movement.

The technologies involved in both types of sensors have seen great improvement over the last few years, and most sensors of each type operate effectively and reliably. Either type of sensor can be effectively integrated with a DDC system. However, there are important differences between the two types of sensors that designers should understand before implementing an occupancy-sensor-based lighting and HVAC control system.

Infrared sensors require line-of-sight sensing. This means that the sensor must always have a "view" of the personnel whose movement is to be sensed. Furthermore, infrared sensors have blind spots—the area of view is usually a checkerboard pattern. Movement, even though it is in the line of sight, that occurs in a blind section of the checkerboard will not be detected. However, the range of an infrared sensor is substantial as long as the line of sight is unobstructed.

Ultrasonic sensors sweep an area with ultrasonic energy so that the line of sight is not required to detect motion. However, ultrasonic sensors can be subject to false alarm by the motion of papers or even air movement at a diffuser. Sensitivity adjustments can reduce these types of problems.

With either type of sensor, it is possible to have triggering occur from external sources. The ultrasonic sensor is susceptible to personnel passing by open doors. Both can be triggered by abrupt changes outside a building when they are installed in a perimeter zone. However, external triggering problems are rare in building areas above the ground level.

Occupancy sensors are usually mounted on the ceiling or high on the wall. Unfortunately, most occupancy sensors are intended to control a lighting circuit directly—without any interface to a DDC system. A typical occupancy sensor comes complete with a power relay to control the lighting circuit and adjustments for the time delay before lights turn off. For a DDC lighting and HVAC control application, the signal from the occupancy sensor is input directly to the DDC system, and the delay logic is included with the other DDC system logic. An occupancy sensor for a DDC system thus does not require a power relay or adjustable-delay circuitry and is less costly. Figure 3-5 shows the comparison between a directly connected occupancy sensor and the DDC system application. Manufacturers have been slow to offer occupancy sensors designed especially for DDC system integration, but the list of products available is now growing rapidly. Infrared occupancy sensors cost $30 to $50 each, and ultrasonic sensors cost slightly more.

Another type of occupancy sensor employs both types of sensing in one unit. This *dual sensor*, as they are typically called, is an excellent product for lighting and HVAC applications. Although these units can be connected in a variety of ways, for HVAC applications it works well to have them connected such that they are not triggered until the infrared sensor detects movement. Thereafter, they provide a signal as long as either sensor detects movement. This approach reduces the false triggering that can occur from ultrasonic sensors, but reduces the risk of terminating occupancy with the occupant(s) still in the area. A further enhancement employs both occupancy sensors and push buttons. The push button enables the occupant to turn off the lights while the room is still occupied, e.g., in order to show slides. It also provides an easy override to establish occupancy in the event that the occupancy sensor fails. If the DDC system receives an input from the push button without a trigger from the occupancy sensor, it means the occupancy sensor has failed. The DDC system can establish occupancy, turn on the lights, and provide an alarm message to the operator that the occupancy sensor is suspect.

Occupancy sensors cost more than the push-button-only approach, but the cost difference can often be justified because the motion-type sensor

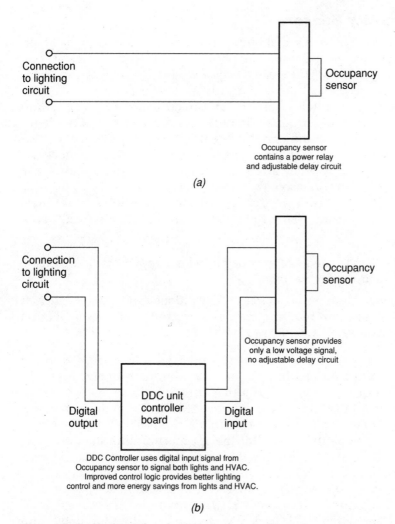

Occupancy sensor
contains a power relay
and adjustable delay circuit

(a)

Occupancy sensor provides
only a low voltage signal,
no adjustable delay circuit

DDC Controller uses digital input signal from
Occupancy sensor to signal both lights and HVAC.
Improved control logic provides better lighting
control and more energy savings from lights and HVAC.

(b)

Figure 3-5. Traditional and DDC system occupancy sensor connections.

offers greater energy savings and less inconvenience to building occupants. It is a nuisance for zones to lapse out of occupancy while still occupied, so most push-button occupancy strategies employ long periods of occupancy each time the button is pushed. If someone comes into an office just to pick up some papers, the zone may remain in the occupied mode for several hours or all day. But an occupancy sensor can automatically return the zone to the unoccupied mode after just a few minutes.

Air-Quality Sensors

Other devices that are growing in popularity are various kinds of air-quality sensors. Such sensors are not required in every zone, but sensors that provide *volatile organic compounds* (VOCs) and CO_2 readings in typical and critical areas can provide useful information to the building control system. An important trend today is clearly toward better regulation of air quality in buildings. Sensors that measure the "quality" of the building environment are becoming economical as new inexpensive technologies are developed to take the measurements. Currently a wide variety of different technologies are used in air-quality sensors, and some are proprietary.

Generally there are two distinctly different questions that must be answered in addressing the issue of air quality in typical buildings:

1. *Ventilation:* Is adequate outside air being supplied to occupants to meet their requirements?
2. *Pollution:* Are internal sources of hazardous contaminants such as off-gassing from building materials or growth of microbiological organisms being adequately regulated?

Unfortunately these very different issues are often lumped together, and as a result, designers and building operators never adequately explore *any* issues that directly affect the indoor air quality of a building. For example, ASHRAE Standard 62-1989, *Ventilation for Acceptable Air Quality*, is the basis for many North American air-quality codes. But the approach in this standard is to combine these two very different issues. It is not difficult to calculate the biological needs of humans for outside-air ventilation. The actual outside-air requirement for humans in most situations is under 2 ft^3/min. The basic ASHRAE standard, however, is 15 ft^3/min. The difference is to deal with the issue of pollution by dilution based on an enormous number of assumptions regarding the character and configuration of the building and its contents. Weaknesses in this approach are well known in the industry, and work continues on developing more precise standards. However, building operators interested in ensuring that good air quality is supplied to their occupants should be aware that in studies of "sick" buildings, the two overwhelmingly *direct* causes have been found to be inadequate mechanical system maintenance, which has permitted the growth of microbiological materials, and the failure of the mechanical system to provide outside-air ventilation at code-required levels. It is therefore not an unreasonable conclusion for designers and operators that keeping mechanical systems clean and well maintained, and providing outside-air ventilation as dictated by the ASHRAE standard, will result in good air quality.

In many instances for today's open-office-plan buildings, an effective means to test and monitor outside-air ventilation is to install a CO_2 sensor on the return air from the building or from individual spaces that are susceptible to problems. This approach will not be effective if the possible air-quality problems are due to air distribution problems within the building or zones. However, if the air distribution system is adequate, the return-air CO_2 readings, when compared with the ambient readings for the building area, provide a good assessment of the building's (or zone's) capacity to expunge the products of the human occupants because CO_2 is a major measurable product of occupants. Information about CO_2 is particularly useful in determining that adequate ventilation is being provided by VAV systems or other air-side economizer systems at minimum outside-air conditions. Attempts have been made to control ventilation directly from such measurements, but this is not wise. Rather, the information can be used to override and adjust the outside airflow upward within limits when a certain threshold is reached.

In recent years another type of sensor has been introduced to the control market that is usually called an *air-quality sensor*. This sensor monitors a number of different contaminants in the air and provides a relative reading of all the contaminants it measures rather than a specific measurement of any one contaminant. These devices are difficult to use effectively because their measurements cannot be compared with accepted standards or other measurements to evaluate the building or even calibrate the sensor. But they do show changes that may affect building occupants which would not otherwise be noted.

Supplying sufficient ventilation air to meet the needs of occupants is the easiest of the problems associated with indoor air quality. Far more difficult is the potential for problems by the contamination of the air by objects or nonhuman organisms. These air-quality problems cannot be directly solved by CO_2 monitoring. For these needs, the industry has begun development of air-quality sensors that are effective in detecting the presence of a range of irritants typically found in buildings. The industry does not yet have sufficient knowledge and experience to determine whether the sensors now available can be effective in this application.

Electric Actuators

Modulating electric actuators have been manufactured for many years, but they have not been widely employed in North American HVAC applications. Until recently, these devices were generally comparatively expensive and were not noted for durability. Electric controls of different manufacturers were often incompatible with one another. The increasing demand for electric control devices that can be interfaced directly with DDC systems appears to be improving the situation. Electric actuators are falling in price and are becoming more standardized on a 0- to 10-V dc control signal. Electric rotating actuators of the type employed are now available for under $100. Electric actuators have several features that make them desirable for HVAC control. Generally, the unit modulates by comparing the modulating voltage from the DDC system with a voltage that depends on the position of the actuator. Unlike pneumatic actuators whose force is related to the output and position, most electric actuators provide constant force (or torque) at all outputs and stop only when the position corresponding to the DDC voltage is reached. This eliminates the uneven movement often associated with pneumatic actuation. Some electric actuators also maintain constant torque whether running or stalled. This feature is particularly useful in applications that require that torque be applied at the end stops to hold dampers fully open or fully closed.

Two-Motor Actuators

A more economical electric actuator for low-torque requirements such as the VAV box damper control employs two small constant-speed clock motors connected together which, as in the actuator above, rotate the shaft collar or clamp through a gear drive. The two-motor type of actuator is interfaced to the DDC system through two digital output points. The two motors run in opposite directions. One motor drives the damper open, the other drives the damper closed. This device is commonly called a *floating-point control actuator*. An example of a two-motor actuator is shown in Fig. 3-6. This type of actuator typically costs under $50 each. The speed of rotation of each motor is constant and is generally geared to be slow—1 to 3 min to rotate fully each direction.

In the past, reliability problems with this type of actuator sometimes resulted when the control system provided continuous power to the controlling motor at the full-open or full-closed position. The

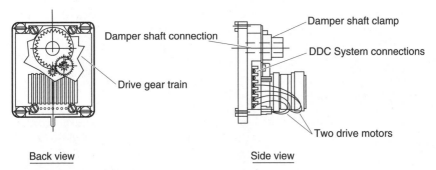

Back view   Side view

Figure 3-6. Two-motor floating-point modulating actuator.

resulting locked-rotor condition appears to have led to abnormal failures of the motor or gear train. Reliability problems also may have resulted from the mechanical switching mechanism that operates the motors. Switches or relays in the control circuit are subject to a great deal of cycling, and unreliable contact closure may have also caused some problems in the past.

Although many of the two-motor actuators are built much more sturdily than they were a few years ago, proper operation by DDC controllers has also contributed to reduced reliability problems. Because DDC systems today typically drive their outputs with solid-state triacs, there are no mechanical components involved in running or switching the motors. To eliminate locked-rotor problems and to provide improved control, several DDC manufacturers have provided special interfaces for the two-motor actuator so that it looks to the operator just like any other analog point. The interface automatically calculates the position of the actuator by keeping track of the time each motor has operated. When the actuator reaches one extreme or the other, the position is automatically reset and the power to the motor is shut off. This approach works well and usually permits two-motor actuators to operate by standard PID controllers just like other analog outputs. Because each motor is shut off at extreme positions, locked-rotor problems are eliminated.

The two-motor controller provides economical modulating control that is adequate for most terminal control applications. These devices are typically configured with slow-speed gear trains because hunting problems can result when the devices are controlled by simple floating-point controllers. The slow speeds may not provide adequate opening or closing speeds in certain circumstances. However, with output controllers that permit PID control of the two-motor actuators, much faster speeds can be controlled with ease, and some two-motor actuators are now becoming available that operate at faster speeds.

What's Needed Next

The quality of typical input/output devices for DDC terminal control is improving while costs are falling. These are good trends. But further improvements are possible, and cooperation with DDC system manufacturers is required to see them through. There remain a number of possible improvements in terminal input/output devices that can increase the cost-effectiveness of DDC systems in the short term.

Modulating Digital Outputs

A variety of electric actuation approaches are offered for reheat valve operation. Those that involve modulation are often more expensive than pneumatic alternates. However, several manufacturers offer small electric two- and three-way valves that employ a very inexpensive method of actuation which could be easily modulated by a DDC terminal controller. These valves are actuated by thermostatic elements similar to the bimetal device that regulates the water temperature in automobiles. But instead of water temperature, these actuators are energized by a small resistance wire that is heated by just a few watts of power. The power requirement is small enough that the valves can be powered directly from the triacs that drive digital output points in most DDC terminal controllers. While these valves are generally used as two-position valves, they can be modulated by pulse-width modulation of the digital output point to which they are connected. These valves cost less than many pneumatic-actuated reheat valves and have a good record for reliability.

Similarly, electric reheat coils can be modulated by pulse-width modulation of a digital output signal to solid-state power relays. Figure 3-7 shows a typical connection for such modulation devices connected to a single DDC digital output.

To make effective use of such modulation techniques, DDC products require the capacity for operators to configure digital output points to "look" like analog points. This is already done by manufacturers who configure two-motor electric actuators to look like a single analog output. Unfortunately, these configurations are not very flexi-

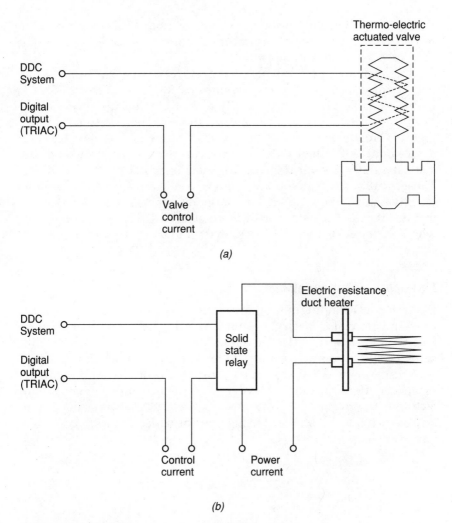

Figure 3-7. Modulating outputs using digital points.

ble now. Rather, they are typically incorporated into dedicated controller outputs that are generally unusable for any other purpose. As was pointed out in Chap. 2, developing products to be more flexible should be the goal of every DDC system manufacturer today. Flexibility in points configuration is extremely helpful to operators. With a configuration option that permits a system to be easily configured with digital outputs operating as analog points, the devices shown in Fig. 3-6 could be operated with standard PID loop control to provide effective modulating terminal control despite the fact that they are digital output points.

DDC system suppliers and manufacturers appear at times to be poorly informed about interface options for their systems. This is understandable since they generally do not manufacture many of the input/output devices employed by their systems. However, if a greater effort were made to seek out effective and economical I/O device options for their DDC systems, controls companies would find a substantially greater demand for their terminal control products. Those manufacturers who are willing to make the effort to provide effective and flexible interfaces to operate special I/O devices in a manner that is consistent with more standard devices are the ones who will dominate the "full" DDC market in the years ahead.

Points to Remember

Electronic DDC input/output devices usually provide far better accuracy and precision than the pneumatic devices that they replace. Electronic devices are also becoming more reliable and cheaper, but many DDC systems do not fully exploit the low-cost devices available. As a result, many DDC system configurations still do not provide the most cost-effective control for high-performance applications. Manufacturers and design engineers need to continue to work together to bring down the cost of effective high-performance control.

4

Programming
Languages

A large part of the enormous improvements in DDC systems that have unfolded in the 1990s is due to advances in the functional capabilities of DDC programming languages. In recent years, the DDC industry has learned much about DDC performance needs and the elements of successful programming languages. A number of ideas have been tried, and many have been very successful. This chapter examines the strategies that the DDC industry has pursued toward DDC control programs and outlines what elements are required for a successful high-performance DDC system application.

Evolution of DDC
Program Languages

In the early days of computer-based building control, most programming languages offered few features and even less flexibility. As a result, the notion developed that the system operator should be more a specialist in computers than in HVAC systems. Many systems were supplied with programs written by the manufacturer at the factory. These programs were provided in a low-level assembly-type language that allowed the operator only a few of what we call "hooks" into the system. For example, typical programs permitted the operator to define occupancy schedules and adjust certain setpoints, but the sequence of control was fixed and could be changed only by recompiling the program, which usually had to be done off-site. Problems that could not be adjusted away with the built-in "hooks" required elaborate schemes to correct. Operators who were very knowledgeable about the operation of the computer could sometimes adjust certain database parameters and fool the system into performing more satisfactorily.

Users became very frustrated by the inflexibility of these early systems. Building control problems that seemed quite simple and straightforward often required elaborate measures and a computer specialist to solve. A flurry of activity took place by manufacturers, users, and the building design community to improve the success of computer-based building control systems. Some of the initial solutions proposed (such as the triservices specification) failed because they attempted to treat the symptoms and not the cause. But when the dust settled early in the 1980s, two new approaches to building control programming were being offered that allowed operators who were not skilled in computers to support building automation systems more effectively than ever before.

Line Programming

One of the approaches to improved programming capabilities offered by some manufacturers was the ability to write sequences of operation in standard line-program formats. Line programming had been employed for many years in the general computing industry. The format of these DDC languages looks very much like the high-level general computing languages (most are similar to BASIC) except certain other functions are added to permit the language to issue start and stop commands, control outputs to PID (proportional, integral, and derivative) algorithms, tie into occupancy schedules, etc. Some languages are compiled and others are interpretive, but all offer similar flexible control capabilities, and they can be easily developed and altered on-site by the system operator. A sample line program in the form that our firm typically employs to calculate the supply air setpoint for simple fan systems is shown in Fig. 4-1.

Some line programming languages contain all the features of powerful high-level languages and include formatting features that make programs written in these languages quite readable. Many also include valuable accessory features such as full-screen editors and on-line error checking that permit operators to view, edit, change, and debug virtually any control sequence quickly and easily. Note, however, that there are wide variations among line programming languages. Some are crude, inflexible, and very difficult to use, although

"CALCULATE MAXIMUM, MINIMUM AND AVERAGE SPACE TEMPS"

STMAX = MAX(ST1,ST2,ST3,ST4,ST5)
STMIN = MIN(ST1,ST2,ST3,ST4,ST5)
STAVE = AVE(ST1,ST2,ST3,ST4,ST5)

"CALCULATE SUPPLY AIR SETPOINT BASED ON AVERAGE SPACE TEMP"

SASP = 65 - 3*(STAVE-STOBJ)

"ADJUST SUPPLY AIR SETPOINT FOR PROJECTED HIGH OUTDOOR TEMP"

SASP = SASP - (PHT-50)/5

"ADJUST SUPPLY AIR SETPOINT FOR COLD DAY MODE OPERATION"

IF CDM = ON THEN SASP = SASP + 2

"ADJUST SUPPLY AIR SETPOINT FOR HIGH OR LOW SPACE TEMPS"

IF STMAX > 74 THEN SASP = SASP - (STMAX-74)*3
IF STMIN < 70 THEN SASP = SASP + (70-STMIN)*3

SASP	is the supply air temperature setpoint
STMAX	is the maximum zone space temperature.
STMIN	is the minimum zone space temperature.
STAVE	is the average zone space temperature.
STOBJ	is the space temperature objective (calculated elsewhere in program).
ST1-ST5	is the space temperatures in areas supplied by air system.
PHT	is the day's projected high outside air temperature (calculated elsewhere in the program).
CDM	is the cold day mode (logically determined elsewhere in the program).

Note: Quotations are comments that are for the benefit of the operator and are ignored by by the program

Figure 4-1. Line program for calculating supply air temperature setpoint.

their suppliers may claim that they provide high-level line programming. As with other features of DDC systems, equals in languages do not exist. Users and consultants should become knowledgeable about the features of any programming language to be supplied with a system before it is purchased.

The primary advantage to line programming rests in its power and flexibility. Line program functions have already proved themselves in solving diverse general computing problems. With the addition of a few special functions for building control, system designers and building operators have the tools they need to develop just about any control sequence(s) for particular HVAC system control requirements. Another advantage of some line programs is that they are self-documenting. When a designer or an operator determines that a program change is necessary, a printout of the program provides an accurate and fairly readable description of the new control sequence. And because line programs are in the form of general computing languages, many operators have already had some experience with this type of language at school or at home.

The primary disadvantage cited for line programming is that some operators have trouble understanding and writing line programs without specific training. Indeed, some line languages have long lists of rules regarding their use, and operators are often frustrated when they cannot easily make occasional program changes. However, those line-based languages with fewer rules which permit the use of comments and special formatting usually are successfully supported by building operators.

Another potential disadvantage of line programs is that software development for typical projects can become time-consuming by requiring entire programs to be rewritten for multiple systems even though they all may operate very much the same. Fortunately, off-line editing features mitigate this problem in the more advanced line-program languages.

Function Block Programming

A second approach taken to improve success with applications software involves refining the preprogrammed approach to give it some additional flexibility in a form structured specifically for typical HVAC applications. The manufacturers who adopted this approach decided to break down the factory-programmed applications into small program blocks that can be linked together and have parameters assigned by the designer or operator. By assembling these preprogrammed blocks in various combinations and providing some flexibility in assigning points and parameters, manufacturers believe they can satisfy most typical DDC applications while maintaining a simple and easy-to-use program format. Figure 4-2 is a sample function block program. This function block provides space-temperature reset of the supply air temperature setpoint of a simple fan system. Some in the industry call this "fill in the blanks" programming because only a limited number of fields in each program block need to be entered.

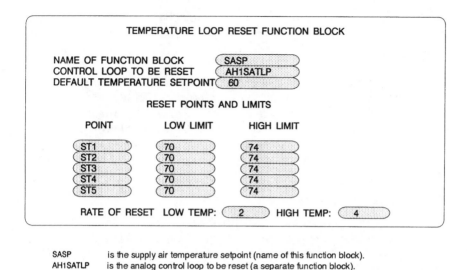

SASP   is the supply air temperature setpoint (name of this function block).
AH1SATLP   is the analog control loop to be reset (a separate function block).
ST1-ST5   is the space temperatures in areas supplied by air system.
()   is the areas of function block that can be changed by the operator.

Figure 4-2. Function block program for supply air temperature reset.

The primary advantage of function block programming is its simplicity in standard HVAC applications. Indeed, if the control sequence happens to call for the exact functions provided by the function blocks included with the system, the programming effort required is very minor. The disadvantage of the function block approach is that whenever sequences are required that do not match available function blocks, the programmer must employ custom blocks or employ blocks for purposes other than those for which they were developed. This usually results in more complicated programs and reduced HVAC system performance.

As a result of the relative advantages and disadvantages of the two programming approaches, line-programming-based systems are usually far more effective in high-performance building applications that require more in-depth control strategies. Systems with function block programming are generally limited to applications employing simple or traditional pneumatic strategies.

To see the differences in the two approaches, consider the programs in Figs. 4-1 and 4-2. The line program in Fig. 4-1 and the reset function block in Fig. 4-2 are both intended to reset the supply air temperature setpoint of a simple fan system as space conditions change. However, the line program in Fig. 4-1 also includes outdoor weather factors and could easily be changed to accommodate different relationships or additional factors—all in this one program. By contrast, the function block cannot implement a number of the factors employed in Fig. 4-1. Function block programs usually do permit linking of blocks for additional factors in calculations. However, linking causes the calculation to be broken into a number of small relationships that do not appear together on a single screen and are therefore difficult for the operator to follow. Programs that require linking function blocks together destroys the simplicity of the system, and simplicity is the main driving force behind function block programming.

DDC Language Trends

For high-performance building control applications, I favor DDC systems that employ line programming. Line programs offer the greater power and expanded functions that are necessary in most high-performance DDC projects. However, manufacturers and users have long understood that there are problems and shortcomings associated with line programming languages in certain applications. Over the years I have worked with users and manufacturers to improve the ease with which line programs may be applied to building control. I promote the concept of *output-oriented* programs wherein every calculation and every command that directly affect an output are installed in one (and only one) program or section of the program. With output-oriented programming, the operator can quickly trace any control path and adjust the program easily when a point or mechanical system is not operating as desired.

In 1988, my firm released a guideline for line-based program languages called the *Operators' Control Language* (OCL). The OCL guide was intended as a functional specification for features that our firm and our clients desired in line-program languages for high-performance applications. A copy of the OCL is included in the Appendix. Building operators are easily trained in the operation of most OCL-based line programs and are able to use many of the advanced features to make their maintenance and troubleshooting duties easier to perform. The reason is that the OCL guide aims at simplifying the use of the control language by consolidating and reducing the rules of application. Improvements are making line programs simpler to apply without compromising the power and flexibility inherent in their architecture.

To improve their range of applications, function-block-based systems must add to their libraries of blocks. But as is true for linking blocks together, large libraries reduce the simplicity of the system. The trend toward high-performance DDC systems is thus eroding the advantages that function block programs offer compared to the more advanced line programs.

New Approaches

The inevitable demand for systems that operate as effectively as operators desire and expect is driving manufacturers to develop innovative approaches to building control applications programming. Several DDC system manufacturers have released DDC systems that employ a new approach to the applications programming language: *graphics programming*. This approach utilizes the powerful graphics capabilities of modern *personal computers* (PCs) to permit the programmer to sketch a flowchart for the control sequence desired. A program in the PC is then employed to translate this sketch into a program (usually written in a high-level line language) which is downloaded to the selected DDC stand-alone panel. An example of a screen containing a graphics program is shown in Fig. 4-3. Note that the entire figure would be built by the operator with system points, variables, and a library of mathematical and logical functions. If the operator desired an additional space temperature for the calculation in Fig. 4-3, it could

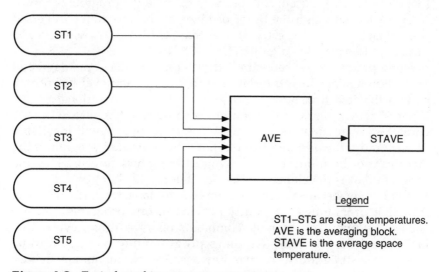

Figure 4-3. Typical graphics program expression.

easily be added by choosing the appropriate system point and sketching a connection to the calculation block.

The idea behind graphics programming is to find a way to provide the power and flexibility of line programming with the simplicity of function block programming. Essentially, the programmer can develop custom function blocks to meet the needs of any particular project application. The idea is enticing, but there are some potential problems with the concept, including the following:

1. *Program execution uncertainty:* Once a programmer has developed a graphic, a translator program has to be employed to convert the picture to an executable program to be downloaded to the appropriate stand-alone panel. Experienced programmers have long realized that there are some special considerations required when writing programs to ensure that they execute as expected. For example, assume the following simple sequence of operation is desired to start a supply and return fan:

If the weekly schedule is on, start the return fan, and after a 30-s delay start the supply fan.

Figure 4-4*a* shows how a line program may be written to execute that logic sequence. If the data point RETURN_FAN is turned on only after the entire block is executed, then Fig. 4-4*a* will execute properly. However, if RETURN_FAN is turned on as soon as that line is executed, it is clear that SUPPLY_FAN may be started an instant after RETURN_FAN is started. A better way to write the program is shown in Fig. 4-4*b*. Although the order of items in the program is reversed from what we might expect, it is clear that SUPPLY_FAN will never be turned on less than 30 s after RETURN_FAN.

While programmers can expect translators to offer some degree of protection against such translation errors, they can never be entirely certain that the translator is not responsible for operational errors. What is a programmer to do if a program does not appear to be executing as pictured in the graphics? The programmer will inevitably be required to inspect and troubleshoot the translated program if the error cannot be found by reviewing the graphics screens. The line program that is developed by the translator is likely to be very difficult to review, because such programs do not have the form or logical flow that programmers typically provide in line programs. Nor are such programs likely to have comments or formatting devices that make them easy to read. Graphics programming may seem to be much more straightforward than line-based software, but any debugging effort can become much more complicated, especially if the operator desires to write more complex control algorithms that make

CONTROL SEQUENCE: If the weekly schedule is on, start the return fan, and after a 30 second dela
start the supply fan.

```
IF WEEKLY_SCHEDULE = ON THEN BEGIN
    DOEVERY 30 SEC
        START RETURN_FAN
        IF RETURN_FAN ON THEN START SUPPLY_FAN
    ENDDO
END
```

(a)

```
IF WEEKLY_SCHEDULE = ON THEN BEGIN
    DOEVERY 30 SEC
        IF RETURN_FAN ON THEN START SUPPLY_FAN
        START RETURN_FAN
    ENDDO
END
```

(b)

WEEKLY_SCHEDULE	is the weekly schedule (set up elsewhere).
DOEVERY 30 SEC	is the do loop that is executed once every 30 seconds.
RETURN_FAN	is the digital output that starts the return fan.
SUPPLY_FAN	is the digital output that operates the supply fan.

Note: The underline character is used to join words to make single terms for point names or
variables. This follows normal programming convention.

Figure 4-4. Line programs to execute simple fan start sequence. (a) Uncertain
execution order; (b) reliable execution order.

the fullest possible use of the energy- and comfort-enhancing capabilities of DDC.

2. *Display limitations:* Another problem with graphics representation of control programs is they are bulky to display. Note that the averaging calculation of Fig. 4-3 requires only a single line to represent in the line program of Fig. 4-1. When several pages or more of graphics are required to represent a control sequence, the sequence can become very difficult to review because the operator cannot look at the whole program at once. I was involved in an experiment in which a line program 50 lines long was translated to a new graphics-program-based DDC system. We found the graphics program required eight screens of graphics representations and seven pages of constants and gains to

duplicate a line program that occupied (with comments) only 1½ pages. The resulting graphics program was far more difficult to understand and review than the original line program.

3. *Operator-interface cost:* While most line-based programming languages can be operated directly or over the phone lines with a simple PC and low-cost software, the complex hardware and software required to create, test, translate, and compile graphics programs can add substantially to the cost of each such terminal. Many users have developed system support mechanisms that include off-site access to the system via telephone modem by the engineer or several operators (at night). These multiple-terminal operational schemes may be very much more costly to implement because simple PCs may not be capable of the performance needed to accomplish graphics programming. Furthermore, the extensive proprietary software required for graphics programming can become costly, and it is possible a copy will have to be purchased for every computer that may be used. Such costs could substantially impact the flexible DDC operating schemes employed by several users.

Current Programming Developments

Despite the fact that the graphics programming packages available today have issues that are of concern to users, the concept represents a serious attempt to satisfy the needs of users. Today, DDC system users and manufacturers alike understand the need to implement DDC systems that are both functional and easy to use. Graphics programming is another serious attempt to provide DDC system operators, whose primary training and knowledge are in the mechanical systems, with the ability to write and adjust high-performance custom DDC applications programs. This represents an important change in operational philosophy from a few years ago when most manufacturers (any many users) believed the DDC system operators should not be permitted direct access to control programs.

While DDC programming features have not surpassed those available to operators in other industries, they have made great strides in recent years to catch up to what is available in other industries. Comparing today's DDC programming features with those only a few years old makes one realize the enormous strides that the industry has made. This rapid development of DDC programming capabilities has moved DDC system capabilities far beyond what designers and users are demanding from their systems in most instances. The current problem is that people are not particularly adept at understanding how effective new digital technologies can be until they have some experience working with them. This makes it difficult to look toward the future with clarity. However, we can look at the issues that need to be resolved in order to continue to improve the success of DDC programming languages, and such a look may provide direction for our efforts.

Combined Line and Graphics Programming

Most high-performance operators continue to believe that line-based programming languages provide the best method to develop DDC programs that execute effective energy- and comfort-enhancing control strategies. It is possible, however, that by combining certain features of both line programming and graphics programming techniques, a format for control programming could result that has advantages over all the programming techniques generally available today.

Earlier in the chapter we showed that line-based programs can be the most effective way to represent many kinds of mathematical and logical expressions because they can make such expressions clearly and concisely. However, a problem with line programs is the lack of clarity in representing major logic sequences. Figure 4-5 shows the logic that might be employed to control a simple fan system. In Fig. 4-5, the logic for the supply fan is very readable in this format because it is simple. The same is generally true for the heating and cooling valves. However, the logic for the mixed-air dampers is not so simple, and therefore it is somewhat difficult to follow even though it is concise.

Function block programming and graphics programming, as they exist today, are also weak in representing system-level logic because they are not concise. In these programs, logic paths are provided, but each component represents a very small portion of the overall program. Therefore, the overall system logic can be represented only after one has pieced together a number of individual blocks. The DDC industry has not yet found an effective means to represent "system"-level logic. This is a shortcoming of every programming format in wide use today.

With the features now generally available in PCs, it may soon be possible to combine line program blocks with graphics representations of system logic to provide a format for improved DDC control

"FAN ON/OFF CONTROL"

```
DOEVERY 1 MIN
   IF OCCUPIED_MODE = ON   OR COOLING_PURGE  = ON   OR
   WARMUP_MODE = ON   THEN START SUPPLY_FAN   ELSE STOP SUPPLY_FAN
ENDDO
```

"HEATING VALVE CONTROL"

```
IF WARMUP_MODE = ON   OR HEATING_REQD = ON   THEN BEGIN
   PID_HTG:SETPOINT = SUPPLY_SETPOINT_CALC
   HEAT_VALVE = PID_HTG
END
ELSE   HEAT_VALVE = 0
```

"COOLING VALVE CONTROL"

```
IF MECH_CLG = ON   THEN BEGIN
   PID_CLG:SETPOINT = SUPPLY_SETPOINT_CALC
   COOL_VALVE = PID_CLG
END
ELSE COOL_VALVE = 0
```

"MIXED AIR DAMPER CONTROL"

```
IF SUPPLY FAN = OFF   THEN MIXED_DAMPER = 0   ELSE
   IF MECH_CLG = ON   THEN IF ENTHALPY_RA = ON   THEN MIXED_DAMPER =
      ELSE MIXED DAMPER = 100   ELSE
         IF COOLING_PURGE = ON THEN MIXED_DAMPER = 100 ELSE BEGIN
            IF MINIMUM_OA < SUPPLY_SETPOINT_CALC THEN
               PID_MAD:SETPOINT = MINIMUM_OA ELSE BEGIN
                  PID_MAD:SETPOINT = SUPPLY_SETPOINT_CALC
                  MIXED_DAMPER = PID_MAD
               END
         END
```

Note: OCCUPIED_MODE, COOLING_PURGE, WARMUP_MODE, HEATING_REQD, MECH_CLG, SUF
LY_SETPOINT_CALC, ENTHALPY_RA, MINIMUM_OA are all variables representing logical decisions or
culations that are not shown in the program. The program represents the logic flow of a simple fan
system that is represented graphically in Fig. 4.6. The underline character is used to join words to
make single terms for point names or variables. This follows normal programming convention.

Figure 4-5. Line program showing control logic for simple fan system.

representations. Figure 4-6 shows the logic for the simple fan control
program of Fig. 4-5 in graphics form. Note that the graphics overview
is effective in representing the system logic. The circumstances under
which the dampers and heating or cooling valves are operated can be
reasonably deduced from the diagram. The labeled rectangular blocks

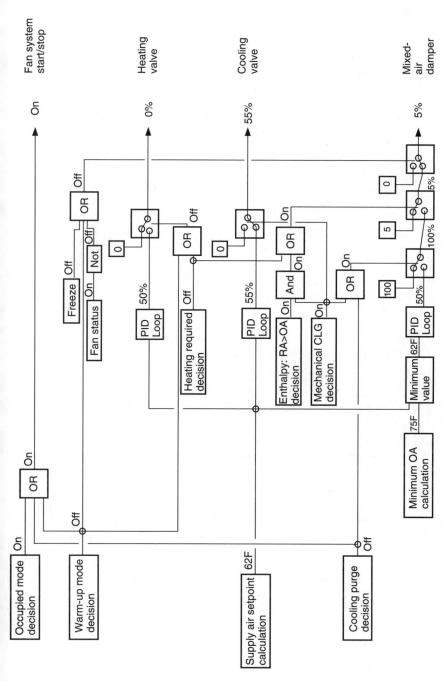

Figure 4-6. System logic for simple fan system.

are line program modules. Imagine that this graphics representation can display real-time result(s) for each line program module, and the current lines of control logic are highlighted in special colors for true or false. Assume further that the contents of any of the line program modules can be called up for review or editing simply by clicking on the chosen module. With such a programming technique the operator could quickly isolate and troubleshoot the exact areas of logic in effect when problems develop. Such a programming scheme as represented in Fig. 4-6 may offer advantages over both line programming and graphics programming while mitigating many of their disadvantages.

Opportunities Ahead

The programming concepts shown above may be a natural evolution as high-performance DDC applications increase, but a wide variety of other options are possible as well. Whatever the next steps in programming languages may be, they will be successful only if they work toward solving the following current problems in high-performance applications:

Concise Representation of Effective DDC Strategies

The key to success with DDCs is not to emulate traditional pneumatic controls, but to use the power and flexibility of DDC systems to provide new, in-depth modes of control that result in enhanced comfort and energy-efficient operation. Such modes of control can be supported only if they are provided in programs that can be understood and diagnosed by the system operators. It is not enough to break the program into such small pieces that the overall concept is difficult to determine because the operator has trouble assembling all the pieces at once. Any language must include representations that show very clearly *both* the overall concept and the calculation or logic pieces that constitute that concept.

Real-time Indications of Program Calculations and Logic Paths

One of the most powerful tools available to a DDC system operator is the ability to watch programs as they execute and to see the results as they are calculated. This programming tool is typically available only with interpretive languages. However it is accomplished, languages must be developed that enhance the operator's ability to view real-time calculations and logic while the DDC system is operating. This feature allows the operator to easily check programs when their operation is suspect.

Few and Simple Rules to Govern the Language

The more rules that govern how a programming language can be applied, the more difficulties the operator has trying to support the programs. Early applications program languages had many rules governing everything from the use of integers and floating-point numbers to the use of mathematics in conditional statements. Manufacturers have done a good job issuing revisions that have simplified language rules for many existing DDC languages, but more needs to be done, and new languages, whether graphics- or line-based, should be as free of restrictive rules as possible.

Low Cost

Improvements in programming languages cannot be permitted to reverse the trend toward lower-cost DDC systems. Manufacturers should consider the enormous market potential for their products when they have combined sufficient function and ease of operation in a package that competes with pneumatics on first cost. "Full" DDC systems are usually 10 to 30 times larger (in terms of system points) than the DDC systems that go into many buildings today. Retrofit opportunities are even greater. Energy costs are now high enough that most building owners can find a very attractive rate of return in an investment of between $1 and $2 per square foot for complete HVAC and lighting control retrofit. If full DDC systems that provide the comfort enhancement and energy reductions of advanced control strategies can be implemented at these costs, the industry will experience an enormous growth in volume over the next few years that will help pay for some of the development requirements.

Points to Remember

The DDC industry has made substantial steps in the last few years to improve the power and flexibility of the control languages supplied with their systems. This is a primary reason that DDC systems are more accepted today than ever before.

The types of control languages commonly available for programming DDC systems today include line-based programming languages, function block programming languages, and graphics programming. Line programming is still the overall choice for most high-performance applications, but each of these approaches has certain advantages for specific applications. Further improvements are still needed, particularly improved representations of "system" logic. When considering programming language improvements, manufacturers need to work toward approaches that permit high-performance control strategies to be represented clearly and concisely and need to provide methods whereby real-time logic paths and calculation results can be displayed as the program is reviewed.

High-performance controls are now being shown to have an enormously positive impact in meeting the expectations of building owners and occupants. To ensure that this higher level of performance can be installed and will be maintained, better-performing and more easily supported applications languages need to be developed. As the power, flexibility, and ease of implementing DDC applications programs grow, the demand for DDC products will grow also.

5

High-Performance HVAC Applications Programs

In Chap. 1, a chilled water distribution network was described to exemplify the challenges and opportunities associated with developing high-performance control strategies for an HVAC application. This chapter will describe how high-performance control can be applied to many other building system control applications. The large variety of specific applications and special considerations make it impossible to cover all possible high-performance HVAC control applications. This chapter is directed toward exploring high-performance control of systems that are widely applied in buildings and have implications for other types of systems as well.

The place to begin in considering high-performance controls is to determine whether the project will benefit from high-performance control. To make this assessment, the designer must evaluate the benefits and assess the costs associated with applying high-performance controls to the application. This determination should be done before the HVAC system and configuration have been determined, because a high-performance control system will affect the HVAC system configuration that is most effective. Public-domain computer simulation programs are usually poor choices for resolving early design issues in a high-performance design process. Most such simulation programs encourage the designer to assume away the real design issues and instead concentrate on calculating total energy use based on relatively low-performance HVAC system technologies. Designers must avoid this pitfall and develop tools that permit a thorough understanding of the heat flows that are likely to exist in various areas of the building under typical conditions.

The tools a designer chooses to assess the high-performance design options should be adequate to accurately depict the building's heat flows and energy needs under actual building operating conditions. This means they must offer dynamic resolution. Hourly simulation is most effective in this regard. But simulation programs need to be straightforward enough that the experienced designer can use the tool directly. Employing design tools that require specially trained people is not an effective means to high-performance system design because this removes the designer from the design process.

To remedy this situation, designers need to look at developing their own design tools for high-performance HVAC systems because such tools are not otherwise available. Design tools must give the designer a means to effectively translate the strategies that evolve from the process to a sound mechanical system design. Firms desiring to develop high-performance designs must do a better job of keeping their designers up to date with the emerging DDC system technologies. My experience is that all too few design firms are willing to make such a commitment. Many firms tend to overstate their knowledge and understanding of advanced technologies. One rarely hears managers from a mechanical-design firm admit that they are not sufficiently current with control technologies to execute high-performance designs, but few have such capabilities. The starting point for any consideration of a high-performance design is for the design firm's principal(s) to commit to acquiring the knowledge and expertise necessary to ensure that the goals of the design will be fully realized.

Features of High-Performance Controls

High-performance control uses advanced technologies to free the designer from the restrictions of steady-state control strategies that dictated design parameters in the era of pneumatic controls. This transition from simple steady-state strategies that were necessitated by the limitations of pneumatic controls to the dynamic control strategies that are now possible with high-performance controls offers important opportunities. Among the most compelling of these opportunities are the substantial reductions in energy use that are possible, the enhancement of comfort levels, and improved indoor air quality.

The methods employed to achieve these results involve the following:

1. *Communication:* High-performance strategies use the improved communication capabilities of modern DDC systems to provide integrated control which permits coordinated operation of HVAC components instead of isolated control by independent control functions, as is basic to pneumatic-based strategies.

2. *Calculation and logic:* High-performance strategies use the advanced logic and computational capacity of DDC systems to anticipate upcoming conditions and more precisely respond to changing conditions.

3. *Instrumentation:* High-performance strategies use the low cost and improved precision of DDC system devices to increase the zoning within a building and to more carefully regulate conditions within each of these smaller zones.

4. *Monitoring:* High-performance strategies use low-cost human interfaces to permit greater operator access and knowledge about the HVAC and building operation.

High-performance HVAC is possible with the new generation of microprocessor-based direct digital control (DDC) systems. With these systems, control is no longer constrained by the input/output relationship established by mechanical devices or electric-resistance bridge circuitry. Compared to pneumatic or electric control systems, microprocessor-based control allows:

Integration: Access to many more input factors to make a control decision or setpoint calculation

Precision: Use of better instrumentation and complex expressions to control according to more precise relationships

As microprocessor-based systems have advanced, many design rules based on earlier control system limitations no longer exist. High-performance dynamic control strategies fit the capabilities of these new microprocessor systems well and improve the energy and comfort performance of most buildings. As the systems have continued to improve, high-performance HVAC concepts are becoming increasingly effective and easy to implement.

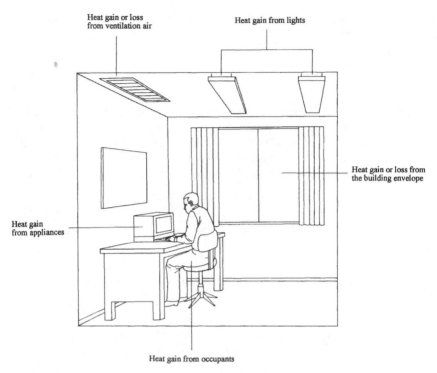

Figure 5-1. Five basic heat flows in typical buildings.

High-Performance Control Is Dynamic

Figure 5-1 illustrates the five basic heat flows in any occupied building. Traditional strategies for HVAC control are based on "steady-state" control, which means that the mechanical system reacts with a heat flow that is equal and opposite to the sum of these five flows. In this manner, a continuous heat-flow equilibrium will be maintained. The equation for this principle is shown in Fig. 5-2.

While maintaining a steady-state condition within spaces of the building is certainly an effective means of maintaining a space-temperature setpoint, steady-state control is usually difficult to maintain because rapidly changing indoor and outdoor conditions often prevent the stable conditions required for steady-state control. The resulting instability of typical steady-state control systems reduces their energy efficiencies and comfort performance.

Still, with the advent of direct digital control systems, most applica-

Provide a continuous heat-flow equilibrium in each zone or

$$\Sigma \, \Delta H = 0$$

where H = changes in enthalpy (heat flow)
Under traditional steady-state control, the HVAC system reacts to
maintain a net heat-flow equilibrium as follows

$$\Delta H_{HVAC} = - \, (\Delta H_{BE} + \Delta H_{L} + \Delta H_{P} + \Delta H_{A} + \Delta H_{V})$$

where ΔH_{BE} = heat flow due to building envelope
$\quad \Delta H_{L}$ = heat flow from lights
$\quad \Delta H_{P}$ = heat flow from people
$\quad \Delta H_{A}$ = heat flow from appliances
$\quad \Delta H_{V}$ = heat flow due to outside air ventilation

Figure 5-2. Steady-state heat-flow balance formulas.

tions software has been patterned after traditional steady-state control. DDC controller modules with reset schedules at the local control-loop level are the primary control. Energy reduction measures are usually accomplished with "add-on" strategies at a different control level and often work counter to the base program. The result can be an unhappy conglomeration of routines working at cross purposes that are hard to understand, adjust, or troubleshoot and that provide neither efficiency nor comfort.

Control strategies based on dynamic control principles have the same overall objective as the steady-state control: providing occupant comfort. But the means of accomplishing this objective is radically different. Dynamic control strategies seek to intelligently manage continuous disequilibrium conditions rather than maintain the steady-state balance of heat flows. Coupled with high-performance control algorithms, this dynamic response will result in much improved system operating efficiencies along with comfort enhancements. A high-performance DDC system, having the capacity to anticipate changing conditions, minimizes the use of energy by manipulating the condition of the building's thermal inertia reservoir with "free" sources of heating or cooling. The dynamic control algorithms in a high-performance system have limiting factors that ensure that the space temperature of each zone is kept within acceptable limits as the thermal inertia conditions of the building are manipulated in anticipation of changing conditions.

The concept of the building as a "thermal flywheel" and the manipulation of the thermal flywheel by outside air, recirculated air, or heat-recovery techniques are not new. But the coordination of these efforts in response to anticipated conditions requires a high-performance DDC system which has only recently become available. Dynamic (time-changing) control is the strategy that underlies the high-performance HVAC concepts discussed in this book. The term *high performance* is used because to be effective, dynamic control requires a level of performance from the HVAC system that is orders of magnitude greater than that traditionally provided.

Dynamic Control Terminology

A number of terms are used in the discussion of high-performance HVAC that are based on dynamic control strategies. We begin by defining the common terms used in this book.

The terms *free heating* and *free cooling* are used to connote energy sources that have no associated marginal costs. For example, recirculated air is a free source of heating, and cool outside air is a free source of cooling during occupancy because the building fans must operate and no additional energy requirement is associated with adjusting the recirculated-to-outside air ratio. Certain kinds of heat-recovery equipment are also sources of free or nearly free heating or cooling.

In addition to free sources of heating and cooling, there are usually low-cost sources. Running the building fans at low speed with 100 percent outside air during the night or before occupancy to precool the building may cost substantially less than operating the chiller at occupancy. Mechanical systems often incorporate heat recovery or thermal storage devices that are sources of low-cost heating or cooling under certain circumstances. For the purposes of this book, all alternate sources are considered with the "free" energy sources. Whether a source should be utilized is an economic question. The designer answers the economic question by developing the high-performance application program such that each is used only when economically advantageous.

For example, a heat-recovery chiller may be viewed as a free source of heating. But if to utilize that free heating, it must operate when it is not required for cooling, then it is not really a source of free heating. Under those circumstances, some other means of providing heating may be more economical. The designer must determine these switchover parameters and include them in the application program.

The *balance temperature* is used to determine when free sources of heating or cooling may be available. Two balance temperatures are used throughout this discussion.

The *heating-balance temperature* is the outdoor temperature at which the heat gain from building occupants, lights, and other internal sources exactly balances the heat loss through the envelope and minimum outside-air ventilation. At any outdoor temperature above the heating-balance temperature, free heating is available, most commonly from return air.

The *cooling-balance temperature* is the outdoor temperature at which maximum outside-air ventilation exactly balances the net heat gain from building occupants, lights, internal sources, and envelope losses. At any outdoor temperature below the cooling-balance temperature, free cooling is available from outside air.

Many buildings are designed to operate between the heating balance temperature and the cooling balance temperature most of the year. This means that free sources of both heating and cooling are usually available to manipulate the thermal condition of the building. However, traditional HVAC and control designs rarely exploit these sources of free or low-cost heating and cooling. High-performance design is the process by which these sources are fully considered, and as we shall see, high-performance control is the manner in which they are exploited in daily building operation.

Getting Ready to Apply High-Performance HVAC Control

High-performance control is a method of managing the more complex control algorithms required to operate buildings dynamically and to fully exploit the sources of free heating and cooling. There are three essential elements of high-performance control that require the designer to look at control design differently. First, setpoint calculations and control decisions in a high-performance applications program are specially organized at a level above the controller level, in order to accept the additional calculation and decision factors necessary for effective control. The local controllers are set up as single-input modules. Resets or control decisions are not made at the controller level. All setpoint calculations and control decisions are made with algorithms and logic in easy-to-read algebraic form in the DDC system's operator's control language (OCL). These setpoints and logic decisions are continuously calculated while controllers are used simply to maintain the current setpoint. Neither controller resets nor cascading controllers are utilized in high-performance controls.

For example, a pneumatic control designer may use nothing more than a two-input controller to establish a mixed-air setpoint and control the dampers based on an outside air temperature reset schedule. In a high-performance control program, the role of the controller modules is simply to modulate the controlling output to maintain a setpoint. The setpoint is a calculated variable whose equation may include a number of other inputs and other variables. Such setpoint calculations enable averages, maximums, minimums, projected conditions, rates of change, and other factors to be included in determining how the output should be operated. The result of the calculation program is a setpoint value. The setpoint may be recalculated and is subject to change every few minutes, but how it will change is entirely determined by the calculation program. The controller module simply overrates the output to maintain the calculated setpoint. In this way control decisions and sequences are made in a program that is integrated with all other HVAC system components. The result is a setpoint calculation program that is comprehensive, readable, and flexible and that offers effective, energy-efficient operation.

In high-performance control schemes, the setpoint calculations and control decisions are fashioned in hierarchical groups. First factors of the setpoints or decisions are established. Then a single expression is used to make the calculation or decision. Figure 5-3 shows a sample mixed-air damper control program in OCL terminology. Note the hierarchy: First the average, maximum, and minimum space temperatures are established; then the factors for each; finally the setpoint is calculated based on these and factors from the predictor modules. The setpoint is directed to the controller which operates the mixed-air dampers when the fan is running. Programs organized in this fashion are easy to start up, debug, and adjust even when they are part of a much larger building control program.

The second essential of high-performance control is its integrated nature which requires that control routines interact together. This ensures that HVAC components work in unison, rather than as a group of independent modules. Dynamic control routines "speak" to one another and are further coordinated by mode setting from weather predictor. The integrated nature of high-performance programs prevents discomfort or energy waste from unnecessary or overlapping operation.

Under traditional local-level control, the chiller and boiler plants are typically activated at fixed outdoor temperatures based on calculated worst-case requirements. These plants operate many more hours than

```
"CALCULATE MAXIMUM, MINIMUM AND AVERAGE SPACE TEMPS"

STMAX = MAX(ST1,ST2,ST3,ST4,ST5)
STMIN = MIN(ST1,ST2,ST3,ST4,ST5)
STAVE = AVE(ST1,ST2,ST3,ST4,ST5)

"CALCULATE SUPPLY AIR SETPOINT BASED ON AVERAGE SPACE TEMP"

SASP = 65 - 3*(STAVE-STOBJ)

"ADJUST SUPPLY AIR SETPOINT FOR PROJECTED HIGH OUTDOOR TEMP"

SASP = SASP - (PHT-50)/5

"ADJUST SUPPLY AIR SETPOINT FOR COLD DAY MODE OPERATION"

IF CDM = ON THEN SASP = SASP + 2

"ADJUST SUPPLY AIR SETPOINT FOR HIGH OR LOW SPACE TEMPS"

IF STMAX >  74 THEN SASP = SASP - (STMAX-74)*3
IF STMIN <  70 THEN SASP = SASP + (70-STMIN)*3

"MODULATING CONTROLLER FOR MIXED AIR DAMPERS EMPLOYS SETPOINT

IF SUPPLY-FAN ON THEN BEGIN
      SETPOINT(MAD-CO) = SASP
      MAD = OUTPUT(MAD-CO)
END
ELSE MAD=0
```

SASP	is the supply air temperature setpoint.
STMAX	is the maximum zone space temperature.
STMIN	is the minimum zone space temperature.
STAVE	is the average zone space temperature.
STOBJ	is the space temperature objective (calculated elsewhere in program).
ST1-ST5	is the space temperatures in areas supplied by air system.
PHT	is the day's projected high outside air temp (calculated elsewhere in the program).
CDM	is the cold day mode (logically determined elsewhere in the program).
MAD-CO	is the mixed air damper controller.
MAD	is the mixed air damper control point (analog output).

Figure 5-3. Mixed-air damper control program.

necessary because worst-case conditions seldom exist. Under dynamic control, equipment operation is the result of a more thorough control algorithm that includes time of day, weather mode, current and projected space, and outdoor conditions, etc. Because a high-performance DDC system has continuous information about the space-temperature and occupancy conditions in every zone of the building, it can operate the boiler and chiller plants much more sparingly but also can more reliably maintain comfort conditions in occupied zones.

The Weather Predictor and HVAC Control

In a building operating dynamically under high-performance DDC, a weather predictor has two principal purposes. First, it switches basic operating modes of the mechanical system between the heating mode and cooling mode. This coordinates the operation of the ventilating system(s) and the heating and cooling plants with the building's space-temperature objective to reduce energy use by maximizing the building as a thermal storage medium. Second, the weather predictor projects the upcoming day's high temperature for use as an anticipatory factor in setpoint calculations.

Why Have a Weather Predictor?

To see the need for anticipatory weather factors in HVAC control, let's look at some common control problems and how the anticipatory feature of a weather predictor can help solve them.

On a cool morning of a warm day, both comfort and energy efficiency can be improved by lowering the mixed-air temperature and the terminal box setpoints to precool the building below the maximum space-temperature limit in anticipation of warmer upcoming conditions. With an anticipatory weather factor, this feature can be automatically incorporated into the setpoint calculations.

A common occurrence in many climates during the spring or fall is cool night and early-morning outside air temperatures. These cool morning temperatures often cause the perimeter radiation system to overreact, resulting in areas of the perimeter overheating as the day warms. This in turn requires early cooling-plant operation and sometimes occupant discomfort. With an anticipatory weather factor, the heating system can be shut down earlier in anticipation of the changing outside conditions. The anticipation eliminates heating overshoot and reduces the hours of required mechanical cooling for a more energy-efficient more comfortable building. Put simply, by anticipating upcoming conditions, the space temperature of a building can be more easily and economically maintained within the comfort range.

What a Weather Predictor Does

The weather predictor performs two functions. It sets the basic operating mode with a digital variable called the *cold-day mode* (often abbreviated CDM), and it projects the day's high temperature with an analog variable called the *projected high temperature* (abbreviated PHT). Let's examine these functions in more detail to see why such variables are valuable in high-performance HVAC applications.

Cold-day Mode

The primary role of the cold-day mode is to reduce overlapping heating and cooling by establishing basic modes for an HVAC system. The cold-day mode is also used to help establish the building space-temperature objective and to coordinate the operation of many HVAC component setpoints. The purpose of this basic mode-setting variable is to attain comfort objectives with sources of free heating or free cooling whenever possible. Only as those free sources, together with the building's thermal inertia, become insufficient to maintain suitable conditions are the primary energy systems activated. This is an overview of how the weather predictor CDM output is used in HVAC control:

Cold-day Mode On

In this mode, the building's space-temperature objective is elevated to a value that depends on the day's projected high temperature. The elevated temperature level in the building is maintained with free sources of heating, such as recirculated heat from people, lights, and appliances. The purpose is to keep the state of the building's thermal inertia reservoir high to minimize the primary heating energy required as colder weather is encountered. In typical CDM operation, the cooling plant is locked out, and heating is used if the building temperatures fall to the minimum comfort level (called the *space-heating setpoint*).

Cold-day Mode off (Warm-day Mode)

In this mode, the building's space-temperature objective is lowered depending on projected high temperature in order to reduce the thermal inertia condition of the building and to delay the requirement for mechanical cooling. Cooling in advance of warm or hot weather is accomplished by depressing the mixed-air temperature and terminal box setpoints to circulate cool morning air through the building. Since the building's thermal reservoir is kept at a low level, it can absorb substantial heat before the mechanical cooling plant is required. In typical *warm-day mode* (WDM) operation, the heating plant is locked out, air systems are controlled to decrease the building thermal inertia by using free cooling, and mechanical cooling is initiated if the building space temperatures rise to a maximum comfort level (called the *space-cooling setpoint*).

Projected High Temperature

Besides establishing basic modes and limits for the mechanical system operation, the weather predictor module calculates another variable that is used to enhance the comfort and efficiency of high-performance HVAC systems. The *projected high temperature* is a factor used in many setpoint calculations to provide an anticipatory factor not available in traditional control strategies. In Fig. 5-3, this factor (called PHT) is present in a typical mixed-air temperature setpoint calculation. Actually, in most dynamic control programs, the CDM and PHT appear either directly or as part of other factors in nearly every major logical decision and setpoint calculation.

Calculating the Two Weather Predictor Factors

Let's now go through the actual process of developing a weather predictor program and examine each submodule in detail.

Outside Air Temperature

A colleague once remarked that one lesson he learned from his involvement in DDC was that input points often don't give the real picture of the conditions. Nowhere is this more evident than with outside-air temperature sensors. Since a continuously reliable outside-air temperature value is an essential input to the weather predictor, it is crucial that the sensor be positioned to be unaffected by sun or other disturbances at all times. I have found only rare occasions when a single outside-air sensor can be adequately shielded from sun and exhaust air such that it provides accurate readings every minute all year round. A solution is to install several outside-air temperature sensors in different locations around the building and to develop a selection program that picks the best value or average of those sensors. The logic flow and sample coding for a simple but effective outdoor temperature selector program are provided in Fig. 5-4.

Day's High and Low
Temperatures and Times

The day's high and low temperatures and the times of their occurrence are helpful historical data used both in the CDM decision and in calculating the PHT. I have found it best to reset the night low temperature every morning just after midnight and to reset the last day's high each day at 10:00 a.m. The logic flow and sample coding for capturing the day's high and low temperatures and the times of their occurrence are shown in Fig. 5-5.

The CDM Decision

As described earlier, the CDM decision is a building-specific determination of outdoor conditions that assesses whether cooling or heating is required to maintain suitable comfort conditions. When the CDM is true, the building is in the heating mode; and when the CDM is false, the building is in the cooling mode.

The CDM is employed instead of just the current outside-air temperature to make the switch from heating to cooling because dynamic control strategies are able to manipulate the building's thermal inertia to make short-term adjustments based on the thermal condition of the building. A building operating under dynamic control strategies does not need to react to the immediate outside-temperature conditions for basic mode changes. Rather it requires information about the general levels of outdoor temperatures that it can expect for the day. Figure 5-6 shows the time-temperature map for 2 days that might cause very different HVAC system operation under traditional controls, but would require very similar energy requirements under dynamic control.

Both day 1 and day 2 contain approximately the same overall average air temperature. The capacity of dynamic control for anticipating upcoming conditions and the use of the building as a thermal storage medium may permit operation on both days without much need for primary heating or cooling-system operation, while a traditional control system may require both heating and cooling systems to operate on day 1.

Because day 1 and day 2 evoke identical limits on the operation of heating and cooling systems for buildings operating dynamically,

Logic Flow

1. Perform this function once every minute

2. Find the maximum and minimum of all outside-air temperature (OAT) sensors.

3. If the range between the maximum and minimum is less than 7°F, the selected outside-air temperature is set equal to the average of the lowest two sensor values.

4. If the range between the maximum and minimum is greater than 7°F, then eliminate each sensor whose absolute difference with the last selected OAT is greater than 5°F.

5. If no sensor is within 5°F of the last selected outside-air temperature, then pick the sensor most likely to read correctly and set the selected outside-air temperature to its value.

Points Legend

OAT   = selected outside-air temperature
OAT1 = outside-air sensor 1
OAT2 = outside-air sensor 2
OAT3 = outside-air sensor 3
A–E   = program variables

Sample Program

```
DOEVERY 1 M

A = MIN ( OAT1 OAT2 OAT3 )
B = MAX ( OAT1 OAT2 OAT3 )
IF ( B-A ) >= 7 THEN GOTO SELECT
ELSE OAT = ( OAT1 + OAT2 + OAT3 - B) / 2
GOTO END
<SELECT>
IF ABS ( OAT - OAT1 ) > 5 THEN C = 0 ELSE C = 1
IF ABS ( OAT - OAT2 ) > 5 THEN D = 0 ELSE D = 1
IF ABS ( OAT - OAT3 ) > 5 THEN E = 0 ELSE E = 1
IF ( C + D + E ) = 0 THEN C = 1
OAT = ( OAT1 * C ) + ( OAT2 * D ) + ( OAT3 * E ) / ( C + D + E )
<END>
ENDDO
```

Figure 5-4. Outside-air temperature select logic flow.

Logic Flow

1. Perform this function once every minute.

2. At 10 a.m. each day, reset the day's high temperature to the present outside-air temperature, and reset the day's high time to 10:00.

3. At 12:30 a.m. each day, reset the day's low temperature to the present outside-air temperature, and reset the day's low time to 00:30.

4. At all other times, capture a new day's high and low temperature and time if the temperature is above (below) the current values of the captured temperatures.

Points Legend

OAT = selected outside-air temperature
DHT = day's high temperature
DHTT = day's high-temperature time
DLT = day's low temperature
DLTT = day's low-temperature time

Sample Program

```
DOEVERY 1 M

IF TIME BETWEEN 10:00 10:05 OR OAT >= DHT THEN
DHT = OAT , DHTT = HOUR
IF TIME BETWEEN 00:30 00:35 OR OAT <= DLT THEN
DLT = OAT , DLTT = HOUR
ENDDO
```

Figure 5-5. Day's high and low temperatures and times logic flow.

some means has to be found to identify them as equivalent and to distinguish them from days of different overall average temperatures for which heating or cooling system operation may be necessary. Obviously current outside-air temperature is not a sufficient delineator, but employing time and temperature together works well. Note that at 10 a.m. on both days the temperature is the same. I have found the temperature at 10 a.m. to be relatively constant for days of similar average temperatures no matter how extreme the highs and lows.

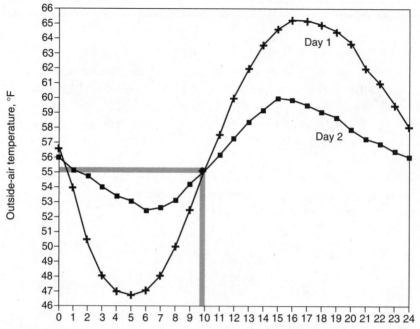

Figure 5-6. Two days with equal average temperature comparison.

Additional work is needed to improve this guideline, but the 10 a.m. rule has worked reasonably well for a number of years, and nothing has yet been found to replace it.

Let us assume that the day's temperatures represented by Fig. 5-6 are the switch-over temperatures. In other words, if a day has a higher average temperature than day 1 or day 2, the cooling system may be required for occupant comfort, and if a day has a lower average temperature than day 1 or day 2, then the heating system may be required for occupant comfort. In this case, the cold-day switch would be 55°F (the 10 a.m. outside-air temperature on days 1 and 2). The *cold-day switch temperature* is defined as the 10 a.m. outside-air temperature for a normally patterned day that separates the need for heating from the need for cooling.

The purpose of the cold-day module is to interpret the current time and temperature in order to determine the correct operating mode. However, one cannot fix the system in one mode and wait until 10 a.m. each day to see if the mode should be switched. The weather predictor solves this problem by constantly checking to see if the current decision is still valid. Each day is separated into three time periods. If the time is between 5 a.m. and 10 a.m., the switch temperature to the CDM depends on the time and on the previous day's high temperature (the higher the previous day's high and the earlier the time, the lower the temperature can drop before entering the cold-day mode).

If the time is between 10 a.m. and 3 p.m., then the module looks at the day's low and the time (the lower the night low and the later the time, the higher the temperature can climb before entering the warm-day mode).

Finally, if the time is between 3 p.m. and 5 a.m., the decision is made only on the basis of the previous day's high temperature. A sample CDM module is presented in Fig. 5-7.

Projected High Temperature

The projected high temperature is an anticipatory factor used in the calculation of setpoints and logic for control parameters much as the immediate outside-air temperature is used in traditional controls.

To be compatible with current DDC languages, current programs to calculate PHTs use relatively simple algorithms that contain historic rise, current time, temperature, and trends. Despite their simplicity, a reasonably accurate PHT algorithm can usually be developed for just about any climate.

The historic data must be very simply stated to be accessed easily in most DDC systems today. The example in Fig. 5-8 employs a simple algorithm that calculates the expected difference between the day's high and the day's low for the day of the year.

Some systems permit the use of lookup tables for this function. The data for this algorithm (or table) are derived from the hourly temperature data in our in-house hourly energy simulation program. The algorithm shown is for Seattle, Washington, in which the average temperature rise between the day's low and the day's high varies quite linearly with respect to the day of the year from about 6°F in the winter to about 25°F in the summer. The data necessary to develop a localized algorithm for other climates can be obtained by analyzing weather bureau summary reports.

If the time is between 8 a.m. and 3 p.m., then the current rate of actual outside-air temperature increase is also factored in the calculation as long as more than 1 h has passed since the time of the day's low temperature. The historical and current-trend factors are then averaged, and a projected high temperature is calculated based on present time and temperature. For the hours between 3 p.m. and 5

Logic Flow

1. Perform this function once every minute.
2. If time is between 5 a.m. and 10 a.m., set up the morning ramp.
3. If time is between 10 a.m. and 3 p.m., set up the daytime ramp.
4. If time is between 3 p.m. and 5 a.m., establish a fixed-ramp value.
5. If the cold-day switch temperature plus the current time-temperature ramp is greater than the current OAT, then the cold-day mode is true. If not, the cold-day mode is false.

Points Legend

CDM = cold-day mode
CDS = cold-day switch temperature—set by the operator (for typical buildings it is 58°F, slightly lower if the building exterior is more efficient than normal and slightly higher if the building is less efficient than normal)
OAT = selected outside-air temperature
DHT = day's high temperature
DLT = day's low temperature
A = local variable

Sample Program

```
DOEVERY 1 M

IF TIME BETWEEN 5:00 10:00 THEN
A = ( CDS − DHT ) * ( 1000 − HOUR ) / 1000
IF TIME BETWEEN 10:00 15:00 THEN
A = ( CDS − DLT ) * ( HOUR − 1000 ) / 2000
IF TIME >= 15:00 OR TIME <= 5:00 THEN
A = ( CDS − DHT ) / 2
IF ( CDS + A ) > OAT THEN CDM = 1 ELSE CDM = 0
ENDDO
```

Figure 5-7. Cold-day decision logic flow.

a.m., the PHT is set equal to the day's high temperature.

The result of this simple calculation can usually be tuned to be more accurate than the weather bureau's forecasts during normal building operating hours, because the PHT is based on real-time data. Sample

PHT modules are shown in Fig. 5-8.

For buildings that operate in evening hours, it is helpful to project upcoming conditions for the evening and night. For this, a *projected low temperature* (PLT) can be calculated from 3:00 p.m. to 5:00 a.m. The PLT algorithm is similar to the PHT calculation.

Putting It All Together

Putting all four of the above submodules together results in a simple weather predictor that is a good starting point for developing high-performance HVAC operational strategies. The weather predictor must be localized for the building's climate. The most important localizing factor is the historical temperature rise factor used in the PHT submodule. This line calculates the expected day's temperature rise (the difference between the day's high and the night's low) for each day of the year. Normally, it will vary between 3 and 35°F, depending on the time of the year and the location. Other factors, such as the ramps for the CDM submodule, may also require adjustment. I recommend that those interested in high-performance HVAC systems begin by setting up and utilizing the weather predictor factors in typical building control programs. Trend logs can be used to tune a weather predictor, and experience with using PHT and CDM is a good start into high-performance operation.

The *space-temperature objective module* can be installed to calculate the level of building temperature that is best suited to meet the projected upcoming conditions. The space-temperature objective module demonstrates how the weather predictor factors are used to create other dynamic control benchmarks. It is a calculation that must be adjusted for comfort requirements of the occupants and limited by the maximum and minimum allowable building space temperatures. A sample space-temperature objective module is shown in Fig. 5-9.

Dynamic Control of Typical Air Systems

The weather predictor module plays an important role in high-performance dynamic control applications. The day's projected high temperature establishes the building's space-temperature objective, which is

Logic Flow

1. Perform this function once every minute.
2. Calculate the day's historical temperature rise (degrees per hour, based on a 10-h linear rise) for the local climate.
3. If the time is after 8 a.m., calculate the actual rate of temperature rise (degrees per hour).
4. If the time is before 8 a.m. or if the day's low has occurred within the last hour, set the actual rise to the historical rise.
5. Set the projected high temperature to the current outside-air temperature plus the average of the historical and actual rise based on a day's high time of 3 p.m. If the time is past 3 p.m., set the PHT to the day's high temperature.

Points Legend

PHT   = projected high temperature
OAT   = selected outside-air temperature
DHT   = day's high temperature
DHTT = day's high-temperature time
DLT   = day's low temperature
DLTT = day's low-temperature time
HRISE = expected temperature rise for 5 a.m. to 3 p.m. (deg/h)
ARISE = actual temperature rise for the day (deg/h)
A–E   = local variables

Sample Program

```
DOEVERY 1 M

IF TIME > 15:00 OR TIME < 5:00 THEN PHT = DHT , GOTO END
B = LIMIT ( ( 1200 − DLTT ) / 500 , 0 , 1 )
HRISE = ( ( 1.3 ) − ABS ( 182 − DAY ) / 180 ) * B * 100
IF TIME < 8:00 OR ( HOUR − DLTT ) < 100 THEN ARISE = HRISE
ELSE ARISE = ( OAT − DLT ) / ( HOUR − DLTT )
PHT = OAT + ( HRISE + ARISE / 200 ) * ( 1500 − HOUR) / 100
<END>
ENDDO
```

Figure 5-8. Projected high-temperature calculation logic flow.

Logic Flow

1. Perform this function once every 2 min.
2. Check to see if the cold-day mode has switched in the last hour. If so, set a ramp to make a 1.5°F switch in the space-temperature objective over a 1-h period.
3. Calculate the space-temperature objective.
4. Set high and low limits on the space-temperature objective.

Points Legend

STOBJ = space-temperature objective
PHT   = projected high temperature
CDM   = cold-day mode

Sample Program

```
DOEVERY 2M

A = 72 − ( PHT − 50 / 20 ) + ( 1.5 * CDM )
STOBJ = LIMIT ( ( A ) ( STOBJ + 0.1 ) ( STOBJ − 0.1 ) )
STOBJ = LIMIT ( STOBJ 74.5 70 )
ENDDO
```

Figure 5-9. Space-temperature objective calculation logic flow.

used as the space-temperature setpoint when free sources of heating or cooling are available to attain it. The weather predictor and other similar programs are known as *global* modules because their outputs—PHT, CDM, and space-temperature objective—are used as inputs to many control modules. Now we shift to discussing specific control modules, and we start with the most important of those modules, the fan control module.

In modern buildings, one or more air systems usually provide ventilation and cooling, through some means of zone control. Heating for perimeter areas may be integrated into the same air system or provided separately. The purpose of traditional air-system controls has been to ensure that each zone receives sufficient cooling or heating to maintain a steady-state equilibrium of heat flows as described earlier. Under dynamic operation, the air system is operated to ensure that each zone temperature is maintained within an acceptable comfort range, but also to utilize free sources of heating (recirculated air) or cooling (outside air). High-performance control manipulates the thermal condition of the building to a state within the comfort range that will minimize the need for primary energy to meet the anticipated upcoming conditions. In warm weather, the space-temperature objective module calculates a setpoint that is near the low limit of the comfort range; and in cold weather, near the high limit of the comfort range. We begin our study of applying dynamic control strategies to air systems by looking at the hardware and software organization required to achieve success with dynamic control operation.

Hardware Organization for Dynamic Control

The DDC system points configuration employed to effect high-performance control should be the simplest configuration that provides the desired control results. Employing more system points or front-end hardware than required for adequate control detracts from the operator's ability to understand and operate the system. Figure 5-10 illustrates the control and monitor points required for a typical constant-volume air system. Humidity sensors would be added in the outside and return air for climates that require enthalpy calculations for the outside/return-air switch. Several additional points would also be required if humidification and dehumidification were included. My experience is that the marginal costs associated with more extensive DDC instrumentation cannot be usually justified by any associated comfort improvements or marginal savings.

Software Organization for Dynamic Control

As discussed earlier in this chapter, high-performance control mandates that all control logic and calculations be done at a level above the controller level. Controllers are configured as simple single-input devices that operate only when called for by the control program and at setpoints calculated by the high-performance control program.

The organization of each specific control module is based on system outputs. All calculations and all logic pertaining to each of the points to be controlled are located in a single module whenever possible.

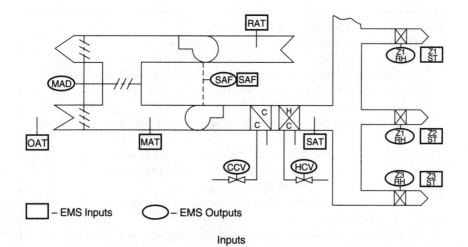

☐ – EMS Inputs   ⬭ – EMS Outputs

Inputs

OAT is the outside air temperature.
SAF is the supply and return air fan status on/off.
RAT is the return air temperature.
MAT is the mixed air temperature.
SAT is the supply air temperature.
Z1ST... ZNST is the zone space temperatures.

Outputs

SAF is the supply and return fans control (start/stop).
MAD is the mixed air damper control (0–100% open).
CCV is the cooling coil valve control (0–100% open).
HCV is the heating coil valve control (0–100% open).
Z1RH ... ZNRH is the zone reheat control (analog/binary as required).

Figure 5-10 Constant-volume terminal reheat system diagram with DDC points.

Grouping all calculations and logic that control each output together facilitates the job of troubleshooting the system. And using a readable operators' control language (OCL) format makes the program easy to adjust, operate, and maintain by the operator and maintenance staff.

High-performance program design for any system is started by outlining the basic architecture of the program, as shown in Fig. 5-11. By following the rules for an output-based program structure, we develop a control module for each of the system outputs as shown in the outline.

1. Supply and return fan on/off control
   a. Run fans continuously during occupied hours.
   b. Start in advance of occupancy to warm building, if necessary.
   c. Operate to maintain the night low limit if required.
   d. Operate for night flush if conditions are suitable.
2. Mixed-air damper control
   a. Calculate a supply air temperature setpoint that maintains minimum ventilation air.
   b. Close dampers if fans are off or mixed-air temperature falls too low, or during warm-up or night low-limit operation.
   c. Open dampers for night flush operation.
   d. Close dampers to minimum position if outside air is warmer than return air.
   e. Operate controller at supply air temperature setpoint at all other times.
3. Cooling coil control
   a. Close cooling coil if fan is off or system is in cold-day mode or night flush is on.
   b. Operate cooling coil if system is in warm-day mode and zone average and maximum temperatures rise above maximum limits.
4. Heating coil controls
   a. Close heating coil if fan is off or system is in warm-day mode or night flush is on.
   b. Operate heating coil if system is in cold-day mode and zone average and minimum temperatures fall below minimum limits.
5. Zone reheat controls (electric reheat)
   a. Stop zone reheat unless system is operating in occupied mode.
   b. Operate zone reheat if system is in the cold-day mode and zone temperature falls below minimum limit during occupied hours.

Figure 5-11 Constant-volume terminal reheat dynamic control program outline.

Dynamic Control of a Simple Constant-Volume Reheat System

Figure 5-12 shows a sample dynamic control program for the system in Fig. 5-10. The program is written according to the OCL guide (see the Appendix). The program for a constant-volume dual-duct system would look very much the same. The sample programs illustrates how the weather factor (cold-day mode) coordinates the operation of pri-

Points Legend

System Points: Refer to Fig. 5.12 for system-interface points.

```
PHT     = projected high temperature (from weather predictor module)
CDM     = cold-day mode (from weather predictor module)
STOBJ   = space-temperature objective (from STOBJ module)
AVEST   = average space temperature
MAXST   = maximum space temperature
MINST   = minimum space temperature
SATSP   = supply air temperature setpoint
HCVSP   = heating temperature setpoint
NFLUSH  = night flush cool-down mode
SCHED   = Scheduled fan operation
OCC     = occupancy time = time of building occupancy
VAC     = vacancy time = time of building vacancy
OPC1    = optimum start response constant (degrees per hour)
OPC2    = optimum start indoor/outdoor response (degrees per degree)
NLLTMP  = night low-limit temperature
A-B     = local variables
```

```
"OCCUPIED SCHEDULE WITH OPTIMUM START"

DOEVERY 1M
AVEST = AVG ( Z1ST Z2ST Z3ST Z4ST Z5ST Z6ST )
MINST = MIN ( Z1ST Z2ST Z3ST Z4ST Z5ST Z6ST )
MAXST = MAX ( Z1ST Z2ST Z3ST Z4ST Z5ST Z6ST )
IF SCHED ON AND A > 120 THEN GOTO FAN.ON
IF SCHED ON THEN A = A + 1 ELSE GOTO FAN.OFF
IF A = 120 AND CDM ON THEN E = 71 - AVEST + E
IF CDM AND ( ( 71 - AVEST + E ) < 120 - ( A / 60 ) * OPC1 )
OR SAFS ON THEN GOTO CALCS ELSE GOTO FAN.ON

"START/STOP OF FAN SAF"

<FAN.OFF>
IF CDM = 1 AND AVEST < ( NLLTMP + ( 2 * SAFS ) ) THEN GOTO FAN.ON
IF NFLUSH = ON THEN GOTO FAN.ON
STOP SAF
A = 0
GOTO CALCS
<FAN.ON>
START SAF
ENDDO
```

Figure 5-12 Constant-volume terminal reheat dynamic control sample program.

mary energy systems. The mixed-air temperature setpoint calculation includes factors designed to anticipate changes in conditions and coordinate the air system to meet the space-temperature objective in the zones. The mixed air is used whenever possible to manipulate the thermal condition of the building in anticipation of upcoming conditions. Primary heating or cooling will be utilized only when this strategy cannot maintain the zones within the allowable comfort range. By inte-

```
"CALCULATE SUPPLY AIR TEMPERATURE SETPOINT"

DOEVERY 5M
SATSP = 3 * STOBJ − ( 3 * AVEST ) − ( PHT / 12 ) + 72
SATSP = SATSP − ( AVEST − B * 10 )
IF CDM ON THEN SATSP = SATSP + 2
IF MINST < 71 THEN SATSP = 71 − MINST * 3 + SATSP
IF MAXST > 74 THEN SATSP = MAXST − 74 * 3 + SATSP
IF SATSP < 50 THEN SATSP = 50
HCVSP = SATSP
IF SCHED ON-FOR < 3H THEN GOTO END
IF SATSP > 0.9 * RAT + ( 0.1 * OAT ) + 1.2
THEN SATSP = 0.9 * RAT + ( 0.1 * OAT ) + 1.2
END
B = AVEST
ENDDO

"CONTROL MIXED AIR DAMPER"

IF SAFS OFF OR MAT < 45 THEN GOTO MAD.CLOSE
IF NFLUSH ON THEN GOTO MAD.OPEN
IF CDM OFF AND OAT > RAT THEN GOTO MAD.MIN
MAD = MADCTL \ GOTO CCV.CTL
[ NOTE: MADCTL IS DAMPER CONTROL LOOP, SET UP SEPARATELY ]
<MAD.MIN>
MAD = 7 \ GOTO CCV.CTL
MAD.OPEN
MAD = 20 \ GOTO CCV.CTL
MAD.CLOSE
MAD = 0

"CONTROL SAF COOLING COIL"

<CCV.CTL>
IF SAFS OFF OR CDM ON OR NFLUSH ON THEN GOTO CCV.CLOSE
IF AVEST < 73.5 THEN GOTO CCV.CLOSE
IF MAXST < 74.5 THEN GOTO CCV.CLOSE
CCV = CCVCTL \ GOTO HCV.CTL
[ NOTE: CCVCTL IS COOLING COIL CONTROL LOOP, SET UP SEPARATELY ]
<CCV.CLOSE>
CCV = 0

"CONTROL HEATING COIL"

<HCV.CTL>
IF SAFS OFF OR CDM OFF OR NFLUSH ON THEN GOTO HCV.CLOSE
IF AVEST > 71 THEN GOTO NCV.CLOSE
IF MINST > 70 THEN GOTO HCV.CLOSE
HCV = HCVCTL \ GOTO NFLUSH.CTL
[ NOTE: HCVCTL IS HEATING OIL CONTROL LOOP, SET UP SEPARATELY ]
<HCV.CLOSE>
HCV = 0
```

Figure 5-12 (*Continued*) Constant-volume terminal reheat dynamic control sample program.

```
"NIGHT FLUSH CONTROL"

IF CDM OFF AND TIME BETWEEN 3:00 8:00 THEN GOTO ST.CALC
ELSE GOTO FLUSH.OFF
<ST.CALC>
IF AVEST > ( 74 - ( 3 * NFLUSH ) ) THEN GOTO OAT.CALC
ELSE GOTO FLUSH.OFF
<OAT.CALC>
IF AVEST - OAT > ( 12 - ( 5 * NFLUSH ) ) THEN GOTO FLUSH.START
ELSE GOTO FLUSH OFF
<FLUSH.START>
START NFLUSH \ GOTO Z1.RH
<FLUSH.OFF>
STOP NFLUSH

"ZONE 1 ELECTRIC REHEAT CONTROL"


DOEVERY 1M

.IF SAFS OFF OR CDM OFF OR SCHED OFF OR SCHED ON-FOR < 2H
THEN GOTO Z1.OFF
IF Z1ST > 70 THEN GOTO Z1.OFF
START Z1RH \ GOTO Z2.RH
<Z1.OFF>
STOP Z1RH
ENDDO

(REPEAT FOR ALL ZONES)
```

Figure 5-12 (*Continued*) Constant-volume terminal reheat dynamic control sample program.

grating the operation of primary energy systems through the weather predictor, heating and cooling will not overlap to waste energy.

Notice that all logic pertaining to the operation of each system output point is located in a single module. Writing custom control modules in the system's general control language has a number of advantages over trying to incorporate a strategy with the "canned" programs provided by the manufacturer. Canned programs such as optimum start usually do not provide the required coordination to other modules or features; to force the mixed-air dampers closed on warmup or to allow the operator to "weigh" the value of each space-temperature sensor in setpoint calculations. Furthermore, canned programs are usually much more difficult to understand and adjust than modules programmed in a powerful and readable OCL.

Although such programs can be read by maintenance staff familiar with the HVAC and control systems, manufacturers must be encouraged to continue their efforts to provide improvements in the function and clarity of control languages. When manufacturers provide flexible and powerful OCLs, the result can be entirely self-documenting programs, whose benefit is that the documentation is automatically updated any time the program is changed.

In some areas of North America, Europe, and Asia, the ability to manipulate a building's thermal inertia through the use of mixed air is limited. However, high-performance dynamic control strategies still result in comfort improvements and savings from improved coordination in the operation of the primary HVAC equipment. These benefits are in addition to the improvements seen during whatever portion of the year the thermal flywheel effect of dynamic control is effective. Furthermore, the same basic strategies and program style can be formulated to operate heat-recovery equipment and thermal storage systems more efficiently, improving the operation of those systems when they are employed.

Dynamic Control of VAV Air Systems

Once installed and adjusted for the specific application, the constant-volume system example of Fig. 5-10 results in a simple and surprisingly energy-efficient HVAC delivery system for many applications. However, variable-volume systems are a better choice when significant variations of heat loading among the zones of an HVAC system exist. A problem with many applications of variable-volume systems is that designers have been encouraged to maintain very low airflows and to run low supply air temperatures to reduce system size and installation costs. The use of low-volume, low-temperature primary air adversely affects comfort, and stratification in occupied areas frequently occurs.

In some instances, series fan-powered terminal boxes are employed to maintain a constant airflow to the space and higher discharge-air temperatures. However, the use of fan-powered boxes reduces the cooling efficiency of the system. The reduction in efficiency is often greater than anticipated because the supply airstream is no longer isolated from the return air.

A New Paradigm in VAV Operation

Combining VAV concepts with high-performance dynamic control strategies is resulting in a new paradigm of energy efficiency and comfort for buildings today. Imagine a building whose systems operate 24 h each day, which does not utilize large night setbacks, and which ensures that minimum ventilation air is provided to each zone any time the zone is occupied. Further, imagine that this building requires only one-half the energy of standard buildings of its type being designed today.

Such high-performance designs are now operating. I call the concept *terminal-regulated air volume* (TRAV). I encourage designers to consider this new approach to VAV design that is possible with the new generation of DDC systems. TRAV designs require the integrated logic of dynamic control to perform effectively.

TRAV designs start with a low-pressure variable-speed VAV system, providing a maximum airflow of about 1 cubic foot per minute or airflow per square root of building area in typical office building applications. The supply fan speed is regulated to meet zone airflow requirements instead of a static pressure setpoint. This approach allows equivalent airflow through the building with substantially lower average fan energy requirements under partial-load conditions. Figure 5-13 is a comparison of a typical variable-speed VAV fan power curve for control by constant static pressure and control by terminal airflow.

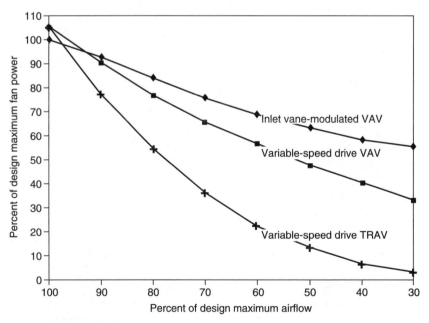

Figure 5-13 Variable-speed fan performance comparison.

Note the low fan energy requirement at the 50 percent airflow range that is possible by replacing static pressure control with flow control.

How Is Terminal-Regulated Air Volume New?

The most popular HVAC system today for commercial buildings is a *variable-air-volume* (VAV) one. Variable-air-volume systems were developed when fully integrated control systems were not available. In VAV systems, regulation of the fan is accomplished independently of the terminal units. The fan is controlled to maintain a constant duct static pressure, and the terminal units provide airflow depending on the zone space temperatures and thermostat settings within field-adjusted limits.

A simple TRAV system outline and the DDC system points interface are shown in Fig. 5-14. Note that the static pressure sensor is placed on the fan discharge. This sensor is employed only as a high-pressure limit and to make certain efficiency calculations. Terminal-regulated air volume (TRAV) is possible today because of the emerging developments in "fully functioning" DDC energy management systems. TRAV is similar to VAV systems in that the volume of airflow is regulated to provide comfort. But TRAV is new because the static pressure regulation characteristic of VAV is replaced with regulation of the central fan directly by the terminal units. Simply stated, terminal-regulated air volume uses advanced applications control software to control the central supply fan based on real-time terminal box airflow requirements rather than to meet a duct static pressure setpoint. This and other system features involving advanced control technologies combine to make TRAV a much more effective and efficient air supply system. A comparison of the features of TRAV and traditional VAV systems is shown in Fig. 5-15.

Fan Regulation

The primary benefit of the TRAV system is its reduced fan horsepower requirements when the system is operating at less than full-flow conditions. Figure 5-13 shows the comparison of the airflow versus fan power for three different system types. The first two are traditional VAV systems, showing flow versus power for VAV with inlet vane control and variable-frequency-drive fan speed control. The

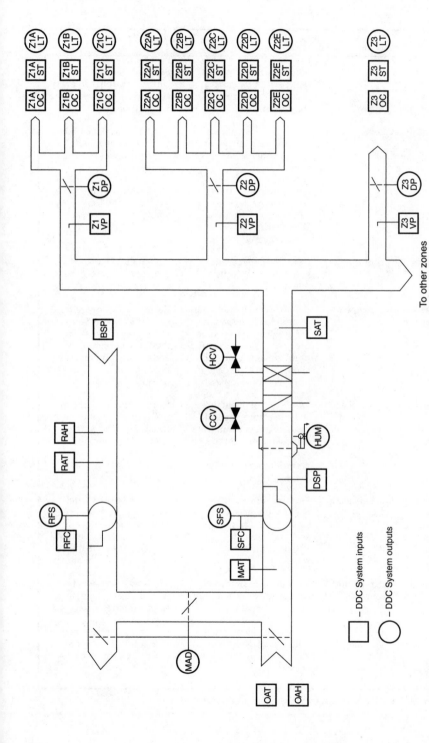

Figure 5-14 Terminal-regulated air volume (TRAV) system diagram with DDC points.

☐ – DDC System inputs

◯ – DDC System outputs

To other zones

113

	VAV	TRAV
1. Fan regulation	Modulate inlet vanes or fan speed to maintain a constant static pressure in supply duct.	Fan control by custom "logic" controller so every zone meets its present airflow setpoint.
2. Zone box control	Local thermostat or sensor controls each box independent of other adjacent zones.	In open office areas, adjacent zones "cooperate" to share the area's heating or cooling load.
3. Air starvation	No coordination if total zone airflow demand exceeds fan capacity. Zones farthest away from fan are starved for air.	Automatic coordination to share capacity limitation among all zones equally.
4. Occupancy control	Operator-entered time schedule or multiple schedules.	Automatic occupancy detection for each zone with occupancy sensors.
5. Low-occupancy operating costs	Relatively high operating costs at low occupancy. Fan horsepower is proportional to airflow, and minimum fan horsepower is about 40 to 50 percent of full-flow horsepower.	Relatively low operating costs at low occupancy. Fan horsepower falls by third power of fan flow, and there is no practical low limit to fan horsepower.
6. Fan volume controller	Loop control maintains static pressure setpoint. Instability can result from control overlap from separate mix-air damper and zone damper control loops.	Fan control by logic controller not PID loop. Instability from control overlap is reduced.
7. Terminal box balancing	Terminal boxes are balanced with mechanical stops that set minimum and maximum flows. Rebalancing is cumbersome.	Terminal boxes are balanced by software applications program. Flow limits can be changed from EMS console easily when distribution of loads changes.
8. Building warm-up and low-limit heating	VAV systems are usually configured as cooling-only systems. Low-limit and warm-up are usually possible at perimeter only—with reheat. Results in long, inefficient warm-up cycles.	TRAV systems switch from central cooling to heating modes easily. Terminal boxes adapt via integrated controls, providing faster and more efficient warm-up cycles.

Figure 5-15 Comparison of VAV and TRAV system features.

third curve shows power versus fan flow for a variable-frequency drive on a fan operating a TRAV control scheme. Note that in a TRAV system, as airflow falls to about 50 percent of maximum, the fan power required falls to less than 15 percent of maximum. The reason is that at a uniform 50 percent airflow condition, the duct static pressure in a TRAV system falls while the zone control dampers remain open. In traditional VAV, the duct static pressure remains constant, while the zone control dampers close to throttle the airflow. The result is the same airflow, but the TRAV control strategy requires far less fan power.

When considering a TRAV system, the designer should carefully investigate the variations in loads for all zones and should design the air duct system accordingly. The cost-effectiveness of a TRAV system will be substantially reduced if duct work is installed such that at low-load conditions, a high duct static is required to provide air to an area such as a computer room whose load may be relatively constant.

Zone Box Control

One significant operating problem of VAV systems is the fact that each terminal box is controlled by a single thermostat, entirely independent of any other box. When a single box supplies air to several small offices, the thermostat in one office controls the flow for all. Or when an open office area is supplied by a number of terminal boxes, calibration differences between the thermostats or small variations in space temperature often cause adjacent boxes to "fight" one another. Both situations often lead to discomfort and air-quality problems. Because TRAV is a fully integrated control approach, the individual boxes act as just another part of the system rather than as isolated devices. In TRAV systems, terminal-box damper or reheat control decisions make use of data from adjacent boxes, or other temperature sensors. In an open office area, the boxes can be configured to work together to "share" the total load. This ensures that small differences in temperature sensor values caused by calibration or other disturbances will not cause the boxes to fight each other. In multiple offices supplied by a single box, a temperature sensor can be placed in each office, and the box can be controlled to meet the combined needs of the spaces. Such an approach is now cost-effective because the DDC system total hardware cost to add space sensors is usually below $50 per point.

Air Starvation

Another nuisance with typical VAV systems is the problem that occurs whenever the zone boxes are calling for more airflow than the central fan can provide. Most VAV systems are designed with a "diversity factor" because the designer assumes the sun will not shine on all sides of the building at once. However, when such a system starts in hot weather, all boxes may call for full cooling, and an air starvation problem exists. During such a condition, the terminal boxes farthest from the central fan usually suffer drastic reductions in airflow, leading to comfort and air-quality problems in the zones they serve.

The integration aspect of TRAV avoids this starvation problem in two ways. First, any time the central fan is running at 100 percent capacity and a call is made for more airflow, a signal is sent to each terminal box calculating routine that reduces each box's airflow setpoint to uniformly distribute the airflow deficit. Such a mechanism leads to more satisfactory precool-mode operation because all areas of the building share the cooling capacity available, and the building is at a more uniform temperature at the conclusion of the cycle.

Second, during occupied hours, the potential problems of air starvation are reduced because of the unique means employed to establish occupancy in each space of a TRAV system. Occupancy sensors shut down airflow to any unoccupied space in the building during high-demand periods, often eliminating the possibility of air deficits entirely.

Occupancy Control

Traditional VAV systems are designed based on an 8- to 12-h occupancy each day and are controlled by a DDC time schedule that usually incorporates some form of optimum start. Once the building is turned over, there is enormous pressure to extend the time schedule hours because employers don't want to discourage employees from working late and because cleaning staff find conditions unsuitable if some air circulation is not provided during their work hours. The reason VAV system operators try to reduce the occupancy schedule duration is because the VAV energy use is relatively high any time the HVAC system is operating, regardless of the low number of building occupants. Furthermore, the associated heating and cooling systems are usually not designed to operate efficiently at low-energy flows.

Extended occupancy hours are designed into the TRAV system to

Occupancy mode	Space-temperature limits Heating	Cooling
Day mode occupied (8 a.m. to 5 p.m.)	72°F	74°F
Night mode occupied	70°F	76°F
Unoccupied	66–69°F depending on OAT	78°F

Figure 5-16 Typical space-temperature limits for TRAV occupancy modes.

provide improved comfort during low-occupancy periods efficiently. A TRAV system uses occupancy sensors in each space to determine the occupancy mode automatically on a zone-by-zone basis. A TRAV system employs a space-temperature setpoint schedule for at least three different conditions: occupied day hours, occupied night hours, and unoccupied hours. A typical setpoint schedule for a TRAV system design appears in Fig. 5-16. It is important to consider that the range between the heating setpoint and cooling setpoint for each category of occupancy is not a dead band. It is the range in which dynamic control strategies employ free sources of heating or cooling to maintain space conditions most efficiently based on projected conditions.

The setpoint schedule shown in Fig. 5-16 usually provides good comfort for occupants and permits efficient operation of the building systems during low-occupancy periods. Our experience has shown that a slightly higher cooling setpoint is perceived by widely scattered occupants to offer about the same comfort level as a lower setpoint in a fully occupied office. Generally, we find occupants are satisfied with wider temperature limits during night and weekend mode hours. We speculate that the reasons are lower expectations and less time spent in the office at these hours.

The temperature schedules shown in Fig. 5-16 can be adjusted for a building (or zones within the building) from the DDC system at any time during its life. Of enormous significance in the TRAV strategy of continuously available occupancy is the concept of constant air circulation. Aside from being the means to control the space temperature during periods of low occupancy, continuous air circulation helps maintain suitable air quality so that occupants are never subjected to a buildup of indoor contaminants. The basic idea of TRAV is to provide at least 20 to 25 percent of the maximum design airflow through the building at all times (requiring about 2 percent of maximum design fan energy) and to introduce amounts of outside air based on outside and space conditions to keep the air quality up to suitable levels at all times. Because of the low fan power costs associated with low-airflow operation, continuous airflow can be maintained in the building at far less cost than with traditional VAV systems, even though the VAV system would be shut down at night and on weekends.

Fan Volume Controller

A concern with VAV static pressure regulation is that the control loops designed to maintain the duct static pressure setpoint do not usually work very well. HVAC designers are so conditioned to working within the limitations of linear control loops that they tend to approach all control problems from the linear-loop perspective.

Unfortunately, many HVAC control requirements like static pressure regulation are not linear. Forcing static pressure regulation into PID loop control is like forcing a square peg into a round hole. The software technician usually finds a way to shave the corners to make it fit, but the result is often unstable in certain situations. VAV fan control problems are exacerbated because the central fan control loop is sandwiched between the mixed-air damper and terminal box control loops, each of which has the capacity to change the air delivery characteristics of the central fan.

Current design practice is to isolate related HVAC system components so that they can be controlled by entirely independent control loops. This usually results in more complicated mechanical systems that have higher first and operating costs. TRAV represents a major technological advance because it employs the integrated capacity of modern DDC systems such that the mechanical system is kept simple. Rather than attempting to isolate related mechanical components, TRAV designs recognize and accommodate the integrated nature of mechanical systems by configuring the DDC system to share information between controllers. This approach results in simpler and less costly mechanical systems as well as smooth, stable operations at all times.

An example of the benefits of integrated control is a supply air system with no return fan. In such a system, the fan inlet static pressure usually increases as the percentage of outside air increases. These variations of inlet static pressure due to the mixing damper position will affect fan flow. But rather than enduring higher first and operating costs of a constant-pressure mixing plenum, TRAV designs employ controls that share information from the mixed-air damper controller to the TRAV fan control module so that fan speed is automatically adjusted to compensate for inlet pressure changes as the mixing damper position changes.

In this way, TRAV designs acknowledge that some HVAC control applications are not well served by independent PID loop controllers. TRAV design relies upon the growing trend in high-performance DDC applications toward custom integrated, logic-based controllers for certain control applications. Such controllers can provide much improved control characteristics for many of the nonlinear components in HVAC systems.

Terminal Box Balancing

In TRAV systems the maximum and minimum airflow limits for each terminal unit are established as a part of the applications software, not as physically installed terminal box limits typical of VAV systems. This TRAV approach permits a more effective building start-up and balancing program and enables the building to be rebalanced with relative ease whenever changes in internal loads require such a measure. The unrestrained operation of the box dampers is necessary to provide steady airflow over a wide range of duct static pressures.

TRAV systems usually employ the software-based airflow limits during daytime occupied hours only. At these times the limits ensure adequate air circulation and outside-air ventilation. During unoccupied or low-occupancy periods, a TRAV system determines airflow setpoints such that air-quality and space-temperature setpoints are maintained at suitable levels with the greatest operating efficiencies.

Heating Cycles for
Building Warm-Up

Although the TRAV system concept does not employ a substantial unoccupied setback, warm-up cycles are important because they must be brief in TRAV designs. Many buildings with VAV systems are difficult to bring up to temperature in cold climates because the interior zones are configured without any capacity for heating. When the space temperatures in the core of the building fall below occupied limits at night or on weekends, the only way they can be raised is with heat from the perimeter which is neither very rapid nor efficient.

TRAV system concepts often utilize a central heating coil for warm-up and low-limit heating during cold weather under low-occupancy conditions. When heating is required, the integrated nature of the controls provides automatic adaptation of each terminal box so that airflow is regulated based on the need for heating in each zone. Because of the continuous airflow characteristic of TRAV, the design temperature rise is small so the heating coil can be designed without a significant pressure differential.

TRAV Operation

It's easy to see why TRAV designs may be well suited for buildings that require efficient occupancy flexibility. The heart of a TRAV system is an integrated, advanced-technology control system, not the mechanical system (which is actually very simple). A wide range of adjustments in TRAV operation can be made easily through the control system console. Because of the dependence on a full-function control system, developing a TRAV system requires a good understanding of DDC capabilities. The control system and how it is configured will ultimately determine whether or not the TRAV system operates successfully.

TRAV is an exciting HVAC system concept that has been shown to be enormously effective. In TRAV applications to date, it has been shown that the control of the central fan by a logic controller from terminal box data can result in a stable system that operates far more efficiently than the VAV system it replaces.

For TRAV designs to be successful, the control system to be employed must be chosen with care. Many DDC systems still cannot provide the full-function control required for TRAV. A number of features that some DDC system manufacturers have adopted, ostensibly to make their equipment easier to program and use, actually have the effect of increasing the difficulty and complexity of applying them to integrated control environments such as a TRAV system.

For commercial buildings that need to provide extended, flexible hours of occupancy, TRAV is an efficient solution to those needs. TRAV is still an emerging HVAC concept, but the effect of TRAV designs is already creating a new paradigm in HVAC performance.

Dynamic Control of
Perimeter Heating Systems

In the discussion of air systems, it was noted that the role of perimeter heating systems is fundamentally changed under high-performance strategies. Now we will examine dynamic control of perimeter heating in detail to see why a new role for perimeter heating systems under dynamic control strategies leads to improvements in energy efficiency and comfort in many buildings.

Why Perimeter Heating Systems Are Required

If one looks at the heat-load characteristics of a typical building, the difference in heating requirements between the interior and perimeter spaces of the building at design conditions is immediately apparent. The difference is the additional heating required to meet the heat loss through the building envelope. Although the interior/perimeter difference has been reduced in modern buildings because the exterior walls are more efficient, it is usually more significant than any other variations among the zones of buildings. Thus, for many years building zones have been divided into interior and perimeter, and a special system is usually employed to meet the heating requirements for the perimeter zones.

Traditional Perimeter Heating System Design Philosophy

The elements of traditional design philosophy regarding perimeter heating systems are based on the steady-state control strategies that have been employed. All these elements deserve close scrutiny in light of the present state of control system capabilities and energy costs.

As discussed earlier, the objective of steady-state control is to maintain a continuous equilibrium within all zones of the building. When it is applied to perimeter heating systems, the result is a heating system concept that envisions balancing the losses through the building envelope so that the exterior areas of the building "look" the same as another interior area for the purposes of the HVAC system. Figure 5-17 is a pictorial representation of this concept with baseboard heating as the perimeter heating source. The idea is to provide just enough heat to offset the building losses. The area served by the perimeter system is often limited to only several feet from the exterior walls. The temperature of the heating water in hydronic systems is usually reset to a schedule that is intended to diminish the heating effect of the perime-

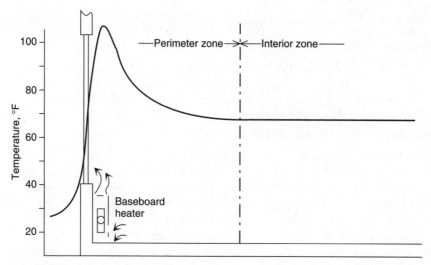

Figure 5-17 Temperature versus distance from exterior wall of baseboard-heated room.

ter-heating system as the outdoor temperature increases, in order to maintain the exterior walls at a temperature close to the space-temperature setpoint.

Perimeter System Operation Changes with Dynamic Control

While the traditional steady-state design approach is a reasonable one, developing adequate controls to achieve the concept is difficult and is almost never accomplished in actual installations. Because interior-air system outlets are often within a few feet of the building perimeter, anything less than ideal control of the perimeter heating can result in significant energy waste from excess perimeter heating that is countered by cooling from the adjacent interior system. Also, additional operating functions are often added to perimeter heating systems that further reduce the energy efficiency of the building. Specifically, perimeter heating systems are widely employed to provide the night low-limit and warm-up heating for buildings. But perimeter systems are not an efficient means of obtaining these functions, as illustrated in Fig. 5-18.

Figure 5-18 is a pictorial representation of the temperatures one might find at various distances from the exterior wall of buildings operating under night unoccupied control. The first temperature curve shows how

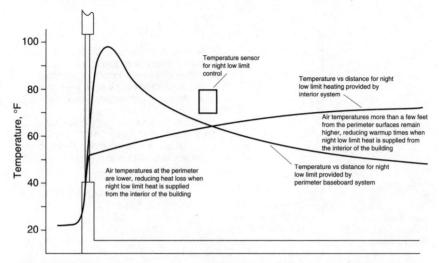

Figure 5-18 Temperature versus distance graph for low-limit heating from perimeter and interior systems.

the air temperature changes with distance from the perimeter walls when night low-limit heating is provided by the perimeter heating system. The second curve shows the variation with night low-limit heating from the interior of the building. Heating by the perimeter system results in a substantially higher building envelope heat loss and much higher energy use. Furthermore, as shown in Fig. 5-18, perimeter night low-limit heating results in a lower interior building space temperature. Interior system low-limit heating maintains the condition of the building's internal thermal mass at higher value through the night, reducing warm-up time the next morning. Buildings designed with high-performance TRAV systems take advantage of this fact to improve energy utilization. Dynamic control strategies employ perimeter systems only as comfort enhancement during occupied hours, not for warm-up or night low-limit heating unless no other means of heating is available.

Dynamic Control Operation During Occupied Hours

To achieve improved energy efficiencies during occupied hours, dynamic control alters the operation of perimeter heating systems during occupied hours also. In traditional hydronic systems, the heat-

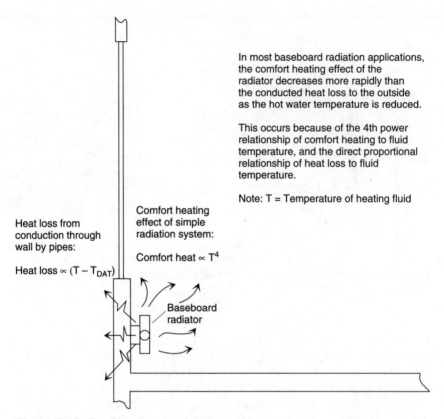

In most baseboard radiation applications, the comfort heating effect of the radiator decreases more rapidly than the conducted heat loss to the outside as the hot water temperature is reduced.

This occurs because of the 4th power relationship of comfort heating to fluid temperature, and the direct proportional relationship of heat loss to fluid temperature.

Note: T = Temperature of heating fluid

Comfort heating effect of simple radiation system:

Comfort heat $\propto T^4$

Heat loss from conduction through wall by pipes:

Heat loss $\propto (T - T_{DAT})$

Baseboard radiator

Figure 5-19 Baseboard system heating and losses.

ing water temperature is reset to make adjustments for envelope heat-loss changes. However, radiant systems run less efficiently at reduced water temperatures. Figure 5-19 shows the relationships between heating water temperature, comfort heating capacity, and unusable heat loss. The radiant, or comfort, effect of a simple radiant system is proportional to the fourth power of water temperature while heat losses are a direct relationship. So, for a simple system, the percentage of unusable heat loss increases as the hot water temperature is reduced. Actual perimeter systems have more complex heat flows than shown, but reduced system efficiency at lower heating water temperatures is common to many types of perimeter systems.

To mitigate this lost efficiency at reduced heating requirements, perimeter systems operating under dynamic control cycle the radiant heating system on and off instead of making the full adjustment by resetting the hot water temperature. This allows the perimeter system to operate at higher efficiencies while reducing its heating effect.

The reason why this approach is successful is because the human body has a built-in averaging capacity that makes it insensitive to short, small changes in the environment. So, under dynamic control, perimeter heating systems are operated continuously only at and near design outdoor temperatures. As the outdoor temperature rises to within 20 to 40°F of the shutoff temperature (depending on the envelope design), the system is cycled. In fact, these outdoor temperatures are those at which perimeter heating systems spend most of their operating hours. A typical perimeter heating system logic flow and sample program for dynamic control operation is shown in Fig. 5-20.

A Dynamic Control Program for Perimeter Heating

The sample program in Fig. 5-20 illustrates the dynamic control objectives for perimeter heating. There are three basic modes for perimeter heat operation. In the warm-day mode, perimeter heating is started only if the building is below the minimum space-temperature requirement by an amount that depends on the projected day's high temperature. Warm-day heating logic is included because the building may cool down over a cold weekend and become occupied on a warmer day.

In cold-day weather, there are two modes of operation possible. If the average and minimum perimeter space temperatures are cool, the perimeter heating runs continuously. As the perimeter spaces warm, the system is cycled to reduce the heating effect. The sample programs do not include a provision for resetting the heating water temperature. In colder climates, the heating water may require some reset in order to meet the wider range of conditions with comfort and efficiency. In these instances, the reset should be limited to about 30°F, and the reset algorithm should include factors for both the outside-air temperature and the average perimeter space temperature. The operation of the cycling program would not be affected by the water temperature reset.

Also note that the sample programs allow perimeter heating only during occupancy. In cold climates, it is sometimes necessary to start the perimeter system somewhat before occupied hours to achieve comfortable conditions at occupancy. However, remember that some

Points Legend

PHT	= projected high temperature (weather predictor module)
CDM	= cold-day mode (from weather predictor module)
Z1ST–Z6ST	= perimeter zone space temperatures
AVEST	= average space temperature (perimeter zones)
MINST	= minimum space temperature (perimeter zones)
OCC	= occupancy time = time of building occupancy
VAC	= vacancy time = time of building vacancy
PHP	= perimeter heating pump on/off control
A–B	= local variables (GCL and UPL)

```
DOEVERY 1M

"CALCULATE AVERAGE AND MINIMUM SPACE TEMPERATURES"
AVEST = AVG ( Z1ST Z2ST Z3ST Z4ST Z5ST Z6ST )
MINST = MIN ( Z1ST Z2ST Z3ST Z4ST Z5ST Z6ST )
"IF BUILDING IS VACANT, STOP PUMP AND GOTO END"
IF ( HOUR < OCC ) OR ( HOUR > VAC ) THEN STOP PHP , GOTO END
"IF THE BUILDING IS IN THE COLD DAY MODE, GO TO HEATING MODULE"
IF CDM = 1 THEN GOTO PERIM-HEAT
"IF THE BUILDING IS IN WARM DAY MODE, START PUMP ON LOW SPACE TEMP"
IF AVEST < ( 71 - ( PHT / 50 ) ) THEN START PHP
IF AVEST > ( 71.5 - ( PHT / 50 ) ) THEN STOP PHP
GOTO END
<PERIM-HEAT>
"IF BUILDING IS COOL, RUN PUMP CONTINUOUSLY"
IF ( AVEST < 72 ) AND ( MINST < 71 ) THEN START PHP , GOTO END
"IF BUILDING AND OAT ARE WARM, KEEP PUMP OFF"
IF ( AVEST > 73 ) AND ( OAT > 50 ) THEN STOP PHP , GOTO END
"OTHERWISE CYCLE PUMP BASED ON OAT, AVEST AND MINST"
IF A <= 0 THEN A = 5 + OAT ELSE A = A - 1
B = 73 - AVEST * 20 + ( 72 - MINST * 10 ) + ( PHP * 10 )
IF B > A AND A > 0 THEN START PHP ELSE STOP PHP
<END>
ENDDO
```

Figure 5-20 Perimeter heating dynamic control sample program.

of the energy released from nearly every type of perimeter system is in the form of radiant energy. I have found in some buildings that occupants of the perimeter are more comfortable when they arrive if the heating system is on, and the space temperature is a degree or two below setpoint, than if the perimeter heating has started earlier, the space is at setpoint, and the perimeter system has cycled off.

The program of Fig. 5-20 requires adjustment and tuning to meet the needs of individual buildings and climates. In the cycle program, note that the length of the cycle decreases as the outdoor-air temperature falls, resulting in shorter on/off cycles with a higher percentage of on time. The length of cycles, the start/stop factors, the switching temperatures, and the need for heating water temperature reset must all be determined individually for each building.

Because high-performance controls operate the interior-air system cooling to achieve the space-temperature objective, which is always higher than the minimum space temperature in cold weather, the interior system cannot mask overheating by the perimeter system. So it is relatively easy to make adjustments to the cycling program module even if an adjacent air system is active.

The program in Fig. 5-20 has also been successfully applied to steam radiation systems. In buildings that have only steam radiators for heating, the use of the three-mode control has resulted in improvements in both energy efficiency and comfort. When the programs in Fig. 5-20 are applied to steam systems, the cycle time should be lengthened, a night low-limit and warm-up module included (if no other means of heating is available for the building), and a space-temperature rate-of-change factor included. Once tuned for a steam radiation system, the dynamic control program provides a substantial improvement in comfort because the cycling reduces the large swings in comfort normally associated with such systems. To prevent water-hammer noises as the system is cycled, I recommend that the steam valve be interfaced to the DDC system with an analog output and a calculation be employed to open the valve at a rate that varies inversely with the time since the last closure.

New Rules for Perimeter Heating System Design

The new approach to perimeter heating, that dynamic control affords, brings about new rules for the design of these systems. We have seen that high-performance control strategies utilize the interior-air system for night low-limit and warm-up, and the perimeter system is cycled as the outdoor temperature rises in order to increase efficiency. I suggest that designers view the perimeter heating system as a "comfort only" system. Its use should not be considered as a contribution to the building's heating requirements, and its operation should not include night low-limit or warm-up operation. Perimeter heating systems are best configured to operate only during occupancy, and their operation should be limited to the times required to maintain satisfactory comfort conditions in the perimeter spaces. Perimeter radiant heating systems should not be operated to balance the heat flow through the building envelope.

Dynamic Control of Boiler and Chiller Plants

Now consider the application of high-performance dynamic control strategies to central plant equipment, the heating and cooling systems that convert primary energy and supply it to those HVAC delivery systems discussed above.

As described previously, high-performance dynamic control strategies bring two new control capabilities to HVAC systems: control based on anticipated upcoming conditions and control that integrates all systems to provide fully coordinated action. Traditional control strategies for establishing the operating hours of central plant equipment are usually very simple. Typically their operation is based on the outside-air temperature and the occupied condition of the building. This results in central plants operating many more hours than necessary because the outside-air temperature switch must be set so that the central plants will operate under any unusual loading conditions that could occur even in mild outside-air temperatures.

With high-performance strategies, custom control algorithms are developed that are aimed at keeping central plants off whenever possible, operating the various central plant equipment only as required by building space conditions (when spaces in the building rise or fall to the limits of comfort). We have seen that when the space conditions are within comfort limits, the HVAC delivery systems are operated to manipulate the thermal mass of the building to delay the need for central plant operation. For example, in the cool morning of a warm day, the mixed-air temperature and terminal box setpoints are lowered to precool the building. The chiller remains off until actual space conditions require mechanical cooling. Thus the full capacity of the building's thermal storage is utilized before primary energy is employed.

Dynamic Control Start/Stop Factors for Central Plants

The objective of dynamic control operation for central plants is to keep the central plant off unless required to maintain suitable space conditions and then to operate the plant as efficiently as possible. To understand the various factors that are used by a high-performance dynamic control program to determine whether the central plant equipment should operate, consider the following factors and how they affect the decision-making process concerning plant operation.

1. *System modes:* As described earlier, every dynamically operated system employs a switch from warm-day mode to cold-day mode to set the basic sequence of the entire HVAC system. In the warm-day mode (cold-day mode off), the boiler is either locked out entirely or allowed to operate only under extraordinary circumstances. The chiller plant is similarly tagged to the cold-day mode. Other mode factors, such as occupied, precool, night low-limit, and warm-up mode, are also employed to regulate the operation of central plant equipment.

2. *Space conditions:* The next most important factor that determines central plant operation is the space condition. Depending on the application, the space condition may include space humidity as well as space temperature. In employing the space condition to control central plant equipment, sufficient logic must be employed to prevent one or more extreme zones from garnering too much authority in determining if a central plant should run. There are several methods employed to avoid this condition. One method is to require that average building conditions as well as one or more individual zones meet certain criteria before central plant equipment is started. Another method is to require that more than a single zone require central plant operation before the plant is started. In massive buildings, the rate of change of space temperature should also be considered, to ensure that the building does not become uncomfortable before the central plant equipment has a chance to react.

3. *Time of day:* If building mode and space conditions signal that the chiller plant is required, but occupancy will end soon, it may be prudent to keep the chiller off if space conditions will not deteriorate significantly in that last hour. Similarly, if space conditions indicate the chiller will be required soon and it is early in the day, then it may be prudent to start the chiller somewhat earlier to ensure that the rapidly warming day does not result in discomfort before the cooling system begins to provide full benefit. For these reasons a time-of-day factor is often employed in the central plant start decision.

Figures 5-21 and 5-22 compare the factors in chiller- and boiler-plant operation between traditional control and dynamic control. Dynamic

Traditional chiller plant start logic requires
  1. OAT ENABLE
  2. FAN(S) ON
Traditional chiller plant stop logic requires
  1. OAT DISABLE OR FAN(S) OFF

Traditional boiler plant start logic requires
  1. OAT ENABLE
Traditional boiler plant stop logic requires
  2. OAT DISABLE

Figure 5-21 Traditional central plant start/stop control factors.

control central plant operation is first enabled by the basic system mode (cold-day or warm-day). We have seen in this chapter that the mode decision is also used to establish the building's space-temperature setpoint and operating setpoints for other HVAC systems. The system mode thus acts to coordinate the manipulation of the building thermal mass condition with the operation of the central plant equipment. In the earlier example of a building operating during the cool morning of a very warm day, we noted that the mixed-air temperature and terminal box setpoints are lowered to precool the building. This will happen only if the system is in the warm-day mode, and in this mode the boiler cannot operate to counter the free cooling process and waste energy. Figure 5-23 shows an example chiller operating program for a building that employs four air-handling systems that are served by a single chiller. Note that the chiller in this instance does not start unless more than one of the systems calls for cooling. Once the basic logic is established, this feature can be changed as required during the tuning process to ensure that the building operates efficiently and comfortably.

Dynamic Sequencing for Central Plants

Once a high-performance program has determined that one of the central plants must operate, the next area of concern is the optimization of the operation of that plant. Optimization usually involves two items:

Dynamic chiller plant start logic requires
SYSTEM MODE ENABLE AND SPACE CONDITION ENABLE
AND OAT ENABLE
   SYSTEM MODE ENABLE requires
     1. System in warm-day mode
     2. Occupied mode OR precool mode
   SPACE CONDITION ENABLE requires
     (Calculate V1 as function of Time of Day)
     (Calculate V2 as function of Time of Day)
     (Calculate V3 as function of Space Temp rate of change)
     1. Average space temperature + V3 > V1
     2. Maximum space temperature + V3 > V2
   OAT ENABLE requires
     1. OAT > C2 (C2 = constant)

Dynamic chiller plant stop logic requires
SYSTEM MODE DISABLE OR SPACE CONDITION DISABLE OR
OAT DISABLE

Dynamic boiler plant start logic requires
SYSTEM MODE ENABLE AND SPACE CONDITION ENABLE
OR FREEZE ENABLE
   SYSTEM MODE ENABLE requires
     1. System in cold-day mode
     2. Occupied mode or warm-up mode or night low-limit mode
   SPACE CONDITION ENABLE requires
     (Calculate V4 as function of Space Temp rate of change)
     1. Average space temperature + V6 < V4
     2. Maximum space temperature + V6 < V5
   FREEZE ENABLE requires
     1. OAT < C4 (C4 is about 0 to 10°F)

Dynamic boiler plant stop logic requires
SYSTEM MODE DISABLE OR SPACE CONDITION DISABLE
AND PLT ENABLE
   PLT ENABLE requires
     1. PLT (projected low temperature) < FREEZE ENABLE TEMP

Figure 5-22 Dynamic control central plant start/stop control factors.

Sample Program

The example is a four-zone system with a single chiller.

Points Legend

Variables:
PHT = projected high temperature (from weather predictor module)
CDM = cold-day mode (from weather predictor module)
Z1ST1–Z4ST6 = zone space temperatures
Z1TAVE–Z4TAVE = average space temperature
Z1TMAX–Z4TMAX = maximum space temperature
Z1TAVEL–Z4TAVEL = last scan average space temperature
CHILLER = chiller status
OCC = occupancy mode
PRC = precool mode
A–B = local variables (GCL)

Logic Flow

1. Mode control
   a. If the weather predictor is in the cold-day mode or the building is not occupied and not in a precool mode, the chiller plant stops and remains off.
   b. Do not start the chiller unless it has been off for at least 30 min.
2. Space conditions
   a. If the highest space temperature in at least two zones is greater than 75°F and the average space temperature of those zones is at least 74°F, then start the chiller.
   b. If the chiller is running, no zone is calling for cooling, and the outside air is less than 65°F, stop the chiller plant.
3. Time conditions
   a. If the time of day is before 12 noon, then lower both the average and maximum zone conditions.
4. Anticycle factor
   a. If the chiller plant is running, keep it running unless the average and maximum zone temperatures fall more than 0.5°F below the maxima required for chiller start and the outside air temperature is below 66°F.

Figure 5-23 Chiller plant dynamic control sample program.

Sample Program in OCL Format

```
DOEVERY 5M
IF ( CDM = 1 ) OR ( ( OCC = 0 ) AND ( PRC = 0 ) ) THEN GOTO STOP
IF ( CHILLER = 0 ) AND ( CHILLER OFF-FOR < 30M ) THEN GOTO END

A = 0
Z1TMAX = MAX ( Z1T1 , Z1T2 , Z1T3 , Z1T4 , Z1T5 , Z1T6 )
Z1TAVE = AVG ( Z1T1 , Z1T2 , Z1T3 , Z1T4 , Z1T5 , Z1T6 )
IF CHILLER = 1 THEN
  B = Z1TAVE + 0.5
ELSE
  B = Z1TAVE + ( HOUR - 1200 ) / 1000 + ( Z1TAVE - Z1TAVEL * 5 )
IF ( Z1TMAX > 75 ) AND ( B > 74 ) THEN A = A + 1
Z1TAVEL = Z1TAVE

Z2TMAX = MAX ( Z2T1 , Z2T2 , Z2T3 , Z2T4 , Z2T5 , Z2T6 )
Z2TAVE = AVG ( Z2T1 , Z2T2 , Z2T3 , Z2T4 , Z2T5 , Z2T6 )
IF CHILLER = 1 THEN
  B = Z2TAVE + 0.5
ELSE
  B = Z2TAVE + ( HOUR - 1200 ) / 1000 + ( Z2TAVE - Z2TAVEL * 5 )
IF ( Z2TMAX > 75 ) AND ( B > 74 ) THEN A = A + 1
Z2TAVEL = Z2TAVE

Z3TMAX = MAX ( Z3T1 , Z3T2 , Z3T3 , Z3T4 , Z3T5 , Z3T6 )
Z3TAVE = AVG ( Z3T1 , Z3T2 , Z3T3 , Z3T4 , Z3T5 , Z3T6 )
IF CHILLER = 1 THEN
  B = Z3TAVE + 0.5
ELSE
  B = Z3TAVE + ( HOUR - 1200 ) / 1000 + ( Z3TAVE - Z3TAVEL * 5 )
IF ( Z3TMAX > 75 ) AND ( B > 74 ) THEN A = A + 1
Z3TAVEL = Z3TAVE

Z4TMAX = MAX ( Z4T1 , Z4T2 , Z4T3 , Z4T4 , Z4T5 , Z4T6 )
Z4TAVE = AVG ( Z4T1 , Z4T2 , Z4T3 , Z4T4 , Z4T5 , Z4T6 )
IF CHILLER = 1 THEN
  B = Z4TAVE + 0.5
ELSE
  B = Z4TAVE + ( HOUR - 1200 ) / 1000 + ( Z4TAVE - Z4TAVEL * 5 )
IF ( Z4TMAX > 75 ) AND ( B > 74 ) THEN A = A + 1
Z4TAVEL = Z4TAVE

IF CHILLER = 0 AND A >= 2 THEN GOTO START
IF CHILLER = 1 AND A = 0 AND OAT < 65 THEN GOTO STOP
GOTO END

<START>
START CHILLER
GOTO END

<STOP>
STOP CHILLER

<END>
```

Figure 5-23 (*Continued*) Chiller plant dynamic control sample program.

1. Determining the sequencing of additional plant equipment

2. Determining the temperature setpoint of the plant output

Because central plant equipment is extremely varied in the configurations employed, it is not possible to generalize as to exactly how the various units should operate. However, what follows is an approach that those of us experienced with dynamic control recommend to improve the efficiency of central plant operation.

Dynamic Control Sequencing of Additional Plant Equipment

If a chiller or boiler plant consists of more than a single chiller (or boiler), some logic must be employed to determine when additional units are started and stopped. The simplest approach is to add and shed units as necessary to maintain the temperature (or pressure) setpoint of the medium. This logic works well when each unit of the plant is small, and it has a very small energy cost overhead associated with starting and stopping each unit.

If each of the central plant units is large and has a substantial energy overhead associated with starting or stopping the unit, another approach is possible with dynamic control. This approach consists of comparing present to upcoming conditions to see if additional plant output is likely in the near future before a unit is added or subtracted. For example, assume a heating plant consists of two large hot-water boilers. If one unit is fully fired and unable to maintain the heating water setpoint, then before the second boiler is fired, a check is made to compare the current outside-air temperature with the projected high temperature (or projected low temperature if the time is past 12:00 noon). If the outside-air temperature is expected to rise enough that the second boiler is not likely to be required later in the day, then it will not be started unless the heating water temperature drops to a level that might adversely affect comfort in the building.

Sometimes the units of the central plant are not the same size or capacity. In this instance a unit would be added that would best suit the expected upcoming conditions as determined by the difference between current and expected *outside-air temperature* (OAT). For example, if in the next 8 h it is expected to become 5°F colder, a smaller boiler might be added than if the OAT drop were expected to be 10°F or more.

These dynamic control operating programs have been applied to the central plant, resulting in excellent energy reductions. However, the logic required is sometimes difficult to apply because many energy management systems today lack the program capacity or versatility to handle such algorithms. Before you decide to develop such an operating program for your central plant equipment, be sure your DDC system can do it. The best way to be certain is to develop an understanding of the language employed by your DDC system and to learn to "think" in that language.

Dynamic Control of Plant Equipment Temperature Setpoint

The second operational control of central plant equipment determines the setpoint of the plant output. Usually this involves determining the best heating or chilled water temperature. The mechanism I recommend is to adjust the temperature as required to ensure that no control valve is more than 100 percent open. However, to be certain that an unstable loop or transient conditions do not result in a single zone artificially controlling the central plant, it may also be necessary to require that the average position of all other valves follow certain other criteria before the supply temperature can be adjusted to meet the requirements of a single valve. For example, if one or more valves are 90 percent open and the average position of all valves is at least 50 percent open, then the temperature setpoint is adjusted at a specified rate until one of those conditions is no longer present. If the maximum valve opening falls to 75 percent or the average of all valves falls to 40 percent, then the temperature setpoint is reversed at a specified rate. In this manner, the central plant setpoint will be adjusted to provide required heating or cooling to all zones without any one zone exerting complete control. Figure 5-24 shows an example program for resetting the chilled water temperature for a chiller supplying four zone coils.

Points to Remember

More than maintaining the building's space-temperature conditions within the comfort range, a building operating under high-performance control is operated dynamically rather than under traditional steady-state control. Dynamic control strategies use free sources of

Sample Program

The example is a four-zone system with a single chiller.

Points Legend

Variables:

Z1CCV–Z4CCV = zone cooling coil-valve position (0 to 100 percent open)

ZVMAX = maximum cooling coil-valve position

ZVAVE = average cooling coil-valve position

CWSP = chilled water temperature setpoint

**Sample Program
in OCL Format**

```
DOEVERY 1M

IF ( CHILLER = 0 ) OR ( CHILLER ON-FOR < 5 M) THEN GOTO STOP
ZVMAX = MAX ( Z1CCV , Z2CCV , Z3CCV , Z4CCV )
ZVAVE = AVG ( Z1CCV , Z2CCV , Z3CCV , Z4CCV )
IF ( ZVMAX > 90 ) AND ( ZVAVE > 50 ) THEN
CWSP = MAX ( 40 , ( CWSP - 0.1 ) )
IF ( ZVMAX < 75 ) AND ( ZVAVE < 40 ) THEN
CWSP = MIN ( 54 , ( CWSP + 0.1 ) )
GOTO END
<STOP>
CWSP = 54
<END>
ENDDO
```

Figure 5-24 Chilled water temperature reset sample program.

heating or cooling to manipulate the building mass to a specific point within the comfort range depending upon projected conditions. In this way the requirements for heating or cooling energy are reduced because the "free" sources assist, either directly or from storage in the thermal mass of the building. High-performance control also entails a far more integrated control system that is capable of maintaining more precise conditions and ensuring that ventilation air is distributed through the building according to actual requirements. Most important is that high-performance HVAC systems such as terminal-regulated air volume (TRAV) are now evolving. Such high-performance mechanical and electric system control concepts can slice the energy use for typical buildings dramatically at the same time as they provide continuous building operation and other valued features. The differences between high-performance and traditional HVAC control concepts pose new challenges to designers. However, the potential benefits not only are worth the effort, but also are required by the significance of those benefits to today's building manager and tenants.

6
Successful High-Performance Design and Project Organization

A profound change is taking place with advanced technologies in the building construction industry. Building design teams have incorporated DDC and other advanced-technology systems in building designs for years, but these systems have rarely played key roles in building design concepts. Now, for the first time in a generation, the industry is experiencing building designs that incorporate new, advanced technologies at the very core of HVAC concepts. In this book we have called this process *high-performance* design.

This emerging notion that advanced technologies should be the basis of designs rather than merely add-on features is what some in the industry see as the *intelligent-building* revolution. Signals are abundant that the revolution is well underway, driven by owners' expectation of more efficient and comfortable buildings. New advanced HVAC concepts based on thermal storage, desiccant dehumidification, and advanced-technology controls are only a few of the new offerings from the design community. Most of these new HVAC concepts are based on sound scientific principles, available now because of recent technological advances. The new HVAC concepts have one important idea in common: They *all* have been advanced to meet projected future needs for comfort conditioning more cost-effectively than conventional HVAC technologies do.

These new system concepts are slowly working their way into new building designs, but employing advanced-technology-based HVAC systems places new demands on the design team. Unless these demands are met, design teams will be less effective in selecting and applying advanced-technology-based designs and will stand to become less competitive. The purpose of this chapter is to outline how advanced technologies can be applied to meet the new demands of intelligent buildings and to suggest a path to follow to ensure that advanced technologies are fairly analyzed and implemented for building design applications.

Advanced Technologies and Intelligent Buildings

According to the Intelligent Building Institute, an *intelligent building* is one which "provides a productive and cost-effective environment through optimization of its four basic elements—structure, systems, services, and management—and the interrelationships between them."

This is a good definition, because it contains the basic objectives necessary for any building construction project to meet future needs. Distilling the definition, I find three characteristics that the development of an advanced-technology-based intelligent building should offer a client:

1. Improved productivity for occupants
2. Optimally cost-effective construction and operation
3. Integrated features that cross design disciplines

These are characteristics that engineers have tried to include in their designs for years. But the limits of traditional HVAC technologies have been in place for so long that many are no longer given adequate consideration in building construction projects. The new technologies have reopened the field to new opportunities. But success in this new environment requires more thorough analyses of design alternatives. Let's begin with a brief overview of the opportunities that have emerged in each of the above-identified areas as a result of increased capabilities of advanced-technology HVAC systems.

Occupant Productivity

A major purpose of any well-designed building is to utilize available technologies to optimize occupant productivity. This is a very important goal because it has been shown that more than 90 percent of the life-cycle costs of typical office buildings are incurred by the people who work in them. Three elements are universally associated with productivity in typical buildings:

1. *Comfort:* There is certainly nothing new about the demand for comfort in buildings. But advanced technologies now offer far better options for ensuring that occupant comfort is effectively maintained. Full direct digital control (DDC) is a new technology that can provide an enormous improvement in building comfort when properly applied. Also, HVAC systems are now becoming available that provide occupants with air distribution and controls at each workstation. We call these systems *individual terminal control* (ITC) units. Tests of prototypes that our firm conducted found that this new technology may offer marked improvements in both comfort and cost-effectiveness.

2. *Environmental quality:* Environmental quality is sometimes linked to comfort, but it should be considered a separate issue by designers. Whereas comfort is a subjective perception, environmental quality is the ability of the environment to meet the physiological requirements of its occupants. Technologies are now available to control the level of many air-based components that affect the environmental quality of a building.

3. *Occupant services:* Occupant services are typically thought of as access to communications, computing, or perhaps secretarial services. While these are all important, forward-looking building designs must ensure that even more occupant service features are incorporated into their designs. The word that fits occupant service features best is *flexibility.* Greater flexibility in building layout is needed so project-oriented firms (like a design firm) can reorganize the office layout to place people working on individual projects together. Greater flexibility in building operation is needed so workers can come to work at any time of day or night or on weekends. Greater flexibility is needed to accommodate heat-load changes that are likely to come in the future from new office equipment, occupant density, appliances, etc.

All these affect productivity. As outlined above, significant recent advances in technology have opened each of these areas to new and more effective approaches that need to be fairly considered for building projects.

Cost-Effective Construction and Operation

The major driving force in the search for new technologies in building construction projects is to develop buildings that are more cost-effective to build and operate. Optimizing cost-effectiveness is not as easy as it used to be. Comparison of construction options must now include utility incentives, alternate sources of funds, scheduling factors, and other criteria. Similarly, building operations comparisons must include far more than utility costs. The complexity of the physical plant affects staffing levels and maintenance costs. How and from where the owner will operate the building are important since economies often exist in operating buildings remotely. All these are factors that owners increasingly wish to see addressed in building designs. Advanced technologies are now available that specifically address these developments.

Integrated Features

As the technology of building systems has advanced, the interrelationship among many far-flung building elements has been recognized. Some of these interrelations, if fully exploited, can offer synergistic advantages in the design and operation of buildings. Full utilization of these features requires an unprecedented effort to coordinate the development of the building design across traditional lines of design discipline. Structural alternatives often change the thermal characteristics of buildings. With advanced digital control, one or more structural alternatives may contribute to reduced HVAC energy costs. There are possible advantages associated with integrating communications, computing, and building control systems. The list of interrelations between different systems that can impact construction or operating cost is nearly endless. To properly exploit each potentially productive interrelation that exists in modern buildings requires a high level of interaction among design team members. Most design teams have long recognized the need to interact more effectively, and in recent years more direct interaction has usually been encouraged. But this interaction is primarily aimed at merely enabling each discipline to complete its design assignments.

Intelligent Buildings and Design

The glamour of an intelligent building cannot be found in its style, the technologies employed, or the uniqueness of its operations. The glamour of a truly intelligent building shows on the bottom line—construction costs, operating costs, and productivity reports. Design team members can improve their abilities to develop intelligent buildings by implementing steps that improve the bottom-line performance of the buildings they design. Specifically, there are two areas that should be continually considered for improvement:

1. Design team organization
2. Tools employed to evaluate design alternatives

HVAC design analysis tools currently revolve around energy simulation programs—many are public-domain programs. Unfortunately, many of the most popular energy simulation programs do not offer the flexibility required to simulate the effects of new technologies on HVAC designs or even to apply existing technologies in new ways to better resolve specific issues. The systems and applications that many popular simulation programs are capable of evaluating are limited, and they don't include recently developed technologies.

This unfortunate circumstance is not easily corrected. The development and the upgrading of widely circulated computer programs are a lengthy process, and the people involved in that process tend to focus on and become experts in the programs themselves rather than remain up to date on industry technologies or applications. Some gap is inevitable between the technologies available to the industry and those upon which the simulation programs are based. The problem is that technologies are now changing sufficiently rapidly to make this gap unacceptable for designers wishing to evaluate the latest HVAC technologies. Today, the most widely used simulation programs are not effective tools to select HVAC systems from advanced-technology alternatives. Nor are such programs generally effective to evaluate the performance of such systems once they are selected.

This is a serious problem because designers need tools that quickly relate alternate design concepts to bottom-line values. To be effective, these evaluations must include effects of integration with components and systems of other disciplines that affect the performance of their designs. Simulation is one of many important tools for HVAC system evaluation. Most of the widely available analytical tools suffer from a similar technological gap. The question is, How does a designer develop and employ an effective evaluation program in this time of rapidly changing technologies?

Intelligent Buildings and Design Evaluation

The moment we acknowledge that a comprehensive evaluation of design alternatives is important and that it is not easily accomplished, we see that this problem is very much like a lot of other problems of the design process, and its solution must be based on a rational plan. I recommend a three-step approach for evaluating advanced-technology HVAC system alternates:

1. Work with each client to develop bottom-line objectives for projects, and agree upon a process that will be used to evaluate design alternatives.

2. Develop in-house computer-based analysis and simulation capabilities to provide many evaluations.

3. Institute follow-up activities to find out how closely the evaluations match actual operation.

These three steps constitute a straightforward and realistic approach to a very real problem. Step 1 is an extremely important feature because it establishes a forum by which the engineer can discuss and understand the owner's needs, desires, and biases concerning each project. Often the owner's view of the requirements of the building construction project is never well understood by the designer, and this can lead to projects that fail to meet owner expectations. This initial contact also provides an opportunity for members of the design team to explain possible values and drawbacks of the advanced technologies that may be considered as design alternates. It is an excellent opportunity for the client and the design team to establish a common ground and to agree on the relative merit of various aspects of the project and how each can be optimized.

The second step—developing in-house evaluation procedures—is controversial. I recommend it because there are no other methods to ensure that new technologies can be accurately evaluated. This approach is opposed by large (primarily public) elements in the industry who would prefer to utilize standard programs for evaluating alternate HVAC designs. I recommend that engineers work to educate these groups on the severe limitations that exist in many of these programs. The current state of the art in programming and computing, together with the extraordinary access to standard computing procedures through organizations like ASHRAE, makes the development of certain in-house simulating capabilities feasible today. I recommend that engineering firms look for programs that can be purchased to be run and modified as necessary by in-house staff. A number of simple but valuable analysis and simulation programs have been developed and are available at reasonable costs. These can be valuable starting points for developing in-house analysis. How much individual engineering firms go toward in-house analysis likely depends on each firm's commitment to employing advanced technologies cost-effectively to solve building design issues.

The third recommended step—follow-up services—is a valuable opportunity for the client and the design firm alike. This service need not be expensive, because if the design is well planned, it is possible to collect the relevant data over phone lines directly into the office computer system for analysis. The resulting analysis is valuable to the client because it can uncover operating or maintenance problems requiring solution. It is valuable for designers because it offers an opportunity to critically analyze their designs to determine how each could be improved to achieve better performance. There is general agreement from building operators that a number of design ideas continue to be employed only because designers never get the feedback from the field as to just how poorly these ideas perform under actual operating conditions.

Because the business of engineering involves applying technologies to designs that optimize performance and cost-effectiveness, design engineers need to pay close attention to the client's bottom line in every project. The intelligent-building revolution requires designers to include a fair and accurate analysis of the newest available technologies in each project. The procedures and tools that are now widely available to the industry for such analyses are not very effective in light of the recent enormous growth of available technologies. Designers must react to this situation and work to develop more effective means so that advanced technologies are employed when they can be effective. The intelligent-building revolution now underway demands this of our industry. It is up those of us in the industry to respond positively and to make the changes required so that our clients get the most cost-effective buildings possible.

New Challenges

Providing effective designs for buildings so that they are implemented and remain true to the design objectives is an enormous challenge to design professionals. Now let's shift our focus to benefits from improvements to the process by which buildings are designed and constructed.

Changing the Design Emphasis

The mechanical designer's activities have been traditionally directed toward very standard mechanical components, air and hydronic systems, chillers, boilers, etc. These mechanical components were all that designers needed to understand to perform their duties. Until very recently, advanced-technology systems such as controls, heat-recovery, or thermal storage systems were seen as add-on items. Manufacturers have been encouraged by designers to package such systems so that they can be selected by the designer much as a chiller or boiler, according to some basic operational criteria.

Today, designers are realizing that integrating advanced-technology systems into the basic mechanical design can simplify the design and provide improved comfort as well as reduced life-cycle costs. The TRAV system concept discussed in Chap. 5 is an example of such an integrated advanced-technology system. In TRAV systems, the high-performance DDC system is central to the entire HVAC concept; it is what makes the concept possible!

The current trend toward employing advanced technologies as the focus of high-performance mechanical system design instead of add-on features is what some in the industry are calling the intelligent-building revolution. The first shots of this revolution were fired only a few years ago, but design firms are already beginning to realize that it marks the beginning of an important new era in the industry. The demand for traditional design services will likely see a great decline when the revolution is over.

While this changing role of advanced-technology systems offers enormous benefits to the success and economy of building designs, it also puts added strains on the design and construction process. Designers need to explore these new strains to the process of building design and construction and need to consider new methods of implementing high-performance systems to meet the needs of the intelligent-building revolution. This chapter suggests an emerging project organization process called *total-involvement engineering* (TIE) that could help relieve the strains and provide greater success with advanced-technology-based designs.

Technology and Building Construction

One of the most challenging demands of the intelligent-building revolution is the problem of translating a design into a mechanical or electric system that meets everyone's expectations for comfort, efficiency, and operability. Traditional building design and construction procedures are a disjointed process in which design and construction are accomplished almost completely independently of each other. This approach has worked in the past because designers and contractors enjoyed a fundamental common understanding of how the components of a building operate. This implicit understanding enabled a designer to specify the components of a mechanical system and then to turn the project over to a contractor to install, commission, and start up. The contractor usually developed an adequate understanding of the system concept just from the arrangement of the components.

Building mechanical systems based on advanced technologies do not usually employ a system concept that can be so easily grasped by a contractor. Designs now employ strategies that are based on building and thermal storage mass, heat recovery, utility rate structures, occupancy patterns, temperature and enthalpy balances, advanced control strategies, and other features whose optimal operation may not be apparent from the mechanical components selected. The designer may have spent a year or more developing the concept. Without a strong presence of the design team during procurement, construction, and start-up, many important criteria of the system concept can be compromised by the failure of others to understand their importance.

Building System Design

Problems in obtaining building systems that work well can often be traced back further than the contractor. Designers find that advanced-technology systems usually require a greater engineering effort than traditional systems. Traditional design team organizational and design fee arrangements often discourage the additional effort required for good advanced-technology system design even though the result may lead to a lower overall project budget and long-term operating improvements. As a result, design processes often develop designs that do not make the best use of advanced technologies and do not meet the owner's expectations for operational efficiency. Figure 6-1 shows a summary of the design and construction issues upon which incorporating advanced-technology systems places added strain. Although most of these issues have been previously cited as weak points of the traditional building construction project organization, their continued influence is retarding the growth of advanced

Issue	Description of problem	Total-involvement-engineering solution
Design team organization	Traditional design teams employ responsibility and communication paths that limit interaction among design team members and with the client. This can result in designs that do not fit together well and fail to meet client expectations.	Communication paths are expanded to accommodate direct communication among team members and client staff. Design team leader manages project, but not communication. Team members are responsible for gathering all information necessary to ensure success of their designs.
Design fees	Establishing design fees as a percentage of the construction cost puts the incentives in the wrong place, and it discourages the use of advanced technologies whose construction costs may be lower but that require greater design efforts.	Design fees are negotiated at the same time as the construction budget is finalized. Client can better evaluate alternatives that have disproportionate impacts on the design and construction budgets.
Procuring advanced-technology systems	Most advanced-technology systems are not functional equals. Selection by price often leads to systems that are not the most cost-effective.	A separate procurement category is established within the construction contract for advanced systems whose selection can be optimized by the designer and client with the use of value-analysis principles.
System commissioning and start-up	Expecting a contractor to grasp the design intent of new technology systems and their interaction with other HVAC systems is unrealistic. Responsibility for commissioning and start-up services is often left for the owner to argue after the fact.	Engineer remains in charge of commissioning and start-up and is responsible for ensuring the system operates as designed.

Figure 6-1. Total-Involvement-Engineering (TIE) improvement guide.

technologies in building construction. Resolution of these issues is imperative if the building construction industry is to provide effective building designs for the next century.

Total-Involvement Engineering

Figure 6-1 lists areas of the building design and construction process in which changes could measurably improve the success of the industry in implementing advanced-technology systems. By adopting and adhering to some new procedures in their design/construction activities, designers and contractors will be much better prepared to deal successfully with the continuing changes in HVAC technologies and building performance requirements.

The integration of these new procedures into the design/construction process is called *total-involvement engineering* because the activities are intended to increase the involvement of the designer in more aspects of the project. The following areas of the design/construction process are recommended for change:

Design: Changes to the design process that enable a greater focus on flexibility and better paths of communication to ensure the owner's expectations for smooth and efficient building operation are met

Procurement: Changes in the procurement process that provide more competitive procurement of advanced-technology systems for which equals do not exist

Commissioning and Start-up: Changes in the commissioning and start-up processes that put the design engineer in charge of start-up and make her or him responsible for obtaining proper operation of the systems he or she has designed

Let's look at each one of these areas in detail and review what procedural changes may be helpful to improve building design success in an environment of advanced-technology system alternatives.

Design Team Organization

Engineers usually try to develop designs that incorporate advanced technologies if they lead to improved efficiency or operations. In practice, however, buildings employing advanced-technology systems very often do not live up to these expectations. Part of the reason for

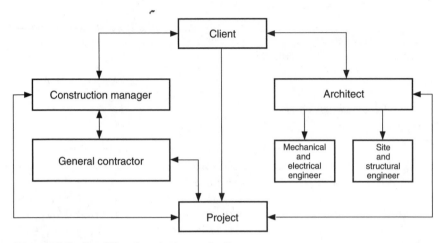

Figure 6-2. Traditional project organization.

difficulties in the design process is the design team organization. Figure 6-2 shows a traditional construction project organization chart that applies to many building construction projects. The project organization is aimed at fixing a minimum number of centers of responsibility and pursuing the various activities in a hierarchical fashion. Thus the design team leader (usually the architect) is ultimately responsible for the mechanical and electrical systems as well as the design activities more closely associated with his or her training. The natural inclination for the one responsible is to control these processes and act as the liaison between the owner and the designer. This mechanism, though cumbersome, usually works satisfactorily as long as traditional technologies are employed. As the state of the technologies involved has advanced, design team leaders have found it increasingly difficult to control all the activities for which they are responsible. Increasingly, the design team has had to decentralize its efforts and encourage designers to work more autonomously.

This decentralization is inevitable and necessary. But because of rigid design team organization, it can cause other problems. Most design teams are organized by delegating responsibilities based on engineering discipline. The design engineer usually becomes focused on just those segments of the project for which he or she is directly responsible. This process frequently discourages the level of communication among design team members required to develop an overall operations plan. An example of this communication failure is illustrated in new buildings that utilize separate control systems for HVAC equipment and lighting and require both to be reprogrammed separately when schedule changes occur.

The development of advanced-technology controls has enabled large owners to operate groups of buildings as units. New buildings are frequently intended to fit into a larger operating scheme. Because the design team often fails to communicate adequately among its members and with the owner, designers often fail to develop their designs such that the completed building will fit efficiently into such an intended operating scheme.

It is usually not possible to expect a designer in any discipline to adequately accommodate her or his design to particular operating schemes unless certain changes are made in the way the design team accepts responsibilities that cross the natural boundaries of design discipline.

Figure 6-3 shows a representation of the project organization many hope will result from the development of TIE approaches. Instead of employing a hierarchical organizational approach, in total-involvement engineering the team concept can be extended to all associated with the building construction project. TIE also emphasizes a strong focus on communication between each member and with the project. The organization represented in Fig. 6-3 is intended to indicate that communication paths at several levels between the participants are encouraged.

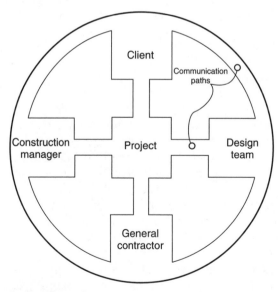

Figure 6-3. Total-Involvement-Engineering organizational wheel.

The responsibility wheel concept of total involvement engineering implies that the major players are sufficiently organized to provide the necessary resources to "contain" any problems or issues that develop during the life of the project. This approach stands in contrast to traditional project organization which assumes that the project will take the proper shape by each party contributing its elements independently. The responsibility wheel concept recognizes that when advanced-technology systems are used, a more coordinated effort by the major players is imperative. Improved communication between groups and within each group also permits responsibilities to be assigned jointly to several team members. For example, a requirement that lighting and HVAC operation be coordinated to ensure that they both operate from a single-schedule system is a responsibility that would be assigned jointly to the mechanical and electrical team members.

Owners are sometimes hesitant to participate in the design effort because they fear designers may try to shift responsibility for decision making to them. The real function of the owner's participation is as a source of important information to the design team—so the designers can better understand owner's operational expectations and react to them—not to mitigate the designers' responsibilities. If this point is made clear at the beginning and throughout the process by strong team leadership, owners often find their participation in the design process to be as valuable to the development of their understanding of the new technologies as it is helpful to the design team.

Design Fees

Although nearly everyone in the building construction industry recognizes that establishing design fees as a percentage of the construction cost is not a good practice, the industry continues to use this as the primary mechanism for contracting designers. There are reasons for its popularity, but the practice works against the best interests of the owner. The designer has a strong incentive to see that every dollar of the construction budget is spent and to choose systems that are higher in construction costs relative to their engineering costs. This traditional approach discourages the use of engineering-intensive advanced-technology systems that could offer major cost and operational advantages.

The TIE approach is to establish an overall budget for design, construction, commissioning, and start-up for each element of the project and to remain open to alternatives that may shift funds among those categories. TIE also encourages built-in flexibility to add items to the budget that meet predetermined criteria for life-cycle cost reduction. The designer remains responsible for the budget, but has greater freedom to propose alternatives that may change the allocation of funds within that budget.

Procuring Advanced-Technology Systems

Current procurement practices have been developed to operate in an environment where a large number of products of equal function and quality are available. Unfortunately, this is rarely the case with advanced-technology products. The low-bid process has become a standard procurement practice in the building construction industry. To competitively procure much of what it takes to construct a building, the low-bid process works well. But when it is applied to purchasing advanced-technology systems wherein equality of function does not exist, the simple low-bid process does not work well at all.

There are some stringent statutory and bureaucratic constraints on procurement procedures for public agencies, but there is always a satisfactory way to modify the procurement process within those constraints when the case for doing so is adequately presented. The need for developing more effective procurement procedures cannot be overstated. An effective procurement procedure is perhaps the largest single factor responsible for achieving successful advanced-technology systems at the lowest cost.

The TIE path to more effective procurement is to treat advanced-technology systems as a special category within the general construction contract. Depending on the constraints of the purchasing entity, it is sometimes possible to conduct an entirely separate procurement process for these systems based on a request-for-proposal process. If not, a prequalification process can be initiated to approve specific configurations of various suppliers that could each provide adequate function. Another means is to require that each general contractor submit at least two or three options for each advanced-technology system of the project. These alternates can be evaluated by the design team based on value-analysis principles before an award is made.

Whatever approach is taken to acquire advanced-technology systems, it is absolutely imperative that the designers develop meaningful specifications for the advanced-technology systems to be employed. All too often, the designers themselves are not certain how to create an effective specification. In these cases, if impartial help is not sought, everyone involved in the project pays for a lack of experience. The TIE process encourages designers to become more knowledgeable about the advanced-technology systems they are incorporating in their designs and to become more straightforward with their specifications.

System Commissioning and Start-Up

A continuing problem with traditional building design and construction is the fact that the design engineer has often completed his or her duties and has left the project by the time the design is ready for commissioning and start-up. This is a bad practice because it denies the engineer the important feedback that can be used to make the next project more effective. It is also harmful to the project because contractors, who do not necessarily understand the concepts behind the physical design, are the ones who must make it work. When systems are simple, little harm can come of this practice, but it can be very detrimental when advanced-technology systems are to be employed. Contractors who do not understand the intent of the design often make adjustments that operate contrary to the intent of the design as they try to make the system perform properly.

The TIE solution is to see that the design engineer remains with the project through start-up. The engineer retains responsibility for ensuring the systems are brought to an operating state that meets all the design criteria. This includes conducting the balancing effort to achieve the design flows and setpoints and taking direct charge of the development of all DDC system software that will provide the control sequences to operate the building. It also includes responsibility for providing the owner's staff with the understanding and training required to operate the building as desired.

Advanced technologies are themselves facilitating this approach. In many new buildings it is now possible to balance the air system from the DDC system keyboard. The engineer can adjust the design criteria to meet any changes in heating and cooling loads that have occurred since the original design was done and to balance the air system from the DDC system console without having to set mechanical stops. The same is true for the hydronic systems and many of the other elements of the HVAC system.

This approach can have an enormously positive impact on the design and construction process. It provides an opportunity for designers to improve their abilities by gaining valuable field experience while making their designs work.

Points to Remember

The elements of a new approach to building design and construction called total-involvement engineering have been developed to improve the success of high-performance building designs that employ advanced-technology systems.

It is important that some improvements to the design and construction process be initiated because the industry is under increased pressure to provide improved comfort and environmental quality and to be certain that improved paths for operational flexibility are incorporated in building designs. Advanced technologies can provide these features, but to do so with the greatest economy requires improvements in the way projects are organized.

If the industry does not accept a leadership role in more effectively implementing high-performance designs, it could find itself applying out-of-date technologies to fewer building projects. But if the industry accepts the challenge, it has the opportunity of increasing its markets of both advanced-technology systems and the design and construction services required to implement them in the increasingly accessible world market.

7

Procurement Procedures for High-Performance DDC Systems

How can designers ensure that the high-performance building systems and components incorporated in a modern building are purchased effectively? Many firms experienced with high-performance DDC systems and other high-performance components have developed more effective procurement techniques with some significant successes—the costs are well below the industry average, and the achievements for these systems are far above the industry norm.

Traditional procurement approaches simplify the process for the designer. But such techniques often result in control or other high-technology systems that are not the best value and often do not meet the functional expectations of the client. Furthermore, the problems with traditional low-bid procurement concepts are increasing dramatically as the use and level of technology involved increase. Figure 7-1 shows the increasing role of high-technology systems in our firm's projects. We believe the use of advance technologies in building mechanical systems will be skyrocketing by the turn of the century.

Criticisms of procuring advanced-technology systems with the traditional specification bidding process fit into two basic categories.

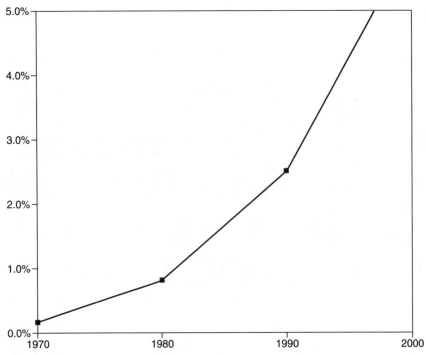

Figure 7-1. Trend toward increase in advanced-technology components as a percent of mechanical system cost.

First, purchasing equipment for which exact equals do not exist is not genuinely competitive and often results in poor value for the client. The cost penalty is potentially enormous. High-performance HVAC control systems that are procured through more competitive means usually meet the applications requirements more effectively and may cost half as much as similar installations (or even less) purchased by the traditional bidding process.

The second basic criticism of purchasing high-performance systems by bidding is that the resulting specifications are generally not clear and require substantial additional design efforts by the successful contractor. By not accepting responsibility for the complete application design of an advanced-technology system, the engineer tends to lose touch with these increasingly important advanced-technology systems. This can be seen with HVAC control systems, which are now as removed from pneumatic controls as office computers are from the typewriters and desktop calculators they have replaced.

Still, a large number of DDC specifications describe function in pneumatic terms.

An engineer who loses touch with certain technologies involved in a design hurts the client because such a designer seldom utilizes that technology to its full potential. Additional problems stem from the ambiguity that results from the designer's lack of understanding of the technology. Many high-technology system specifications actually discourage contractors from responding with low-cost or optimally configured systems.

In this chapter, we suggest ideas for alternative procurement procedures to engineers and owners who are contemplating high-performance systems or components for new or existing building mechanical systems. The techniques discussed are not new and are not the only possible options. They are techniques that firms have used successfully for a number of years. As buildings continue to employ advanced-technology systems, the techniques that engineers and owners develop to purchase, install, and operate these advanced-technology components will increasingly influence the level of success that these systems attain.

Separate Procurement Categories

Designers and owners should consider establishing a separate category at the start of each building construction project for systems and equipment that utilize advance technologies and, because of a lack of equal products, require a special evaluation process for selection. The owner and engineer should then work together to ensure that these systems and components are designed to meet the functional requirements of the project with as high a degree of certainty as possible and that they are purchased as competitively as possible. The following three-step process is suitable for a number of other advanced-technology systems and components that may be integrated into HVAC system designs.

Step 1: What's Needed

The first step in conducting an effective procurement process for advanced-technology systems is determining what features are needed and desired for the system to operate successfully as part of the building's mechanical system.

Function

Building owners, including speculative developers, almost always want more than just the cheapest product. They desire systems that will provide suitable levels of performance at the most reasonable cost. The engineer must determine a part of what constitutes a "suitable level of performance" by determining what is necessary to meet the rigors of the overall mechanical design. However, the owner's concerns should be an important factor in determining what is suitable as well. Design engineers are generally not well acquainted with issues of operability or maintainability—especially as they relate to the rapidly advancing technologies that are incorporated into DDC systems and other energy management devices.

Modern mechanical designs are becoming increasingly complex because they must meet increasingly complex demands for space and energy efficiency, comfort, air quality, and fire and life safety. An engineer may spend months developing and refining the overall operating strategy for an HVAC system, and the engineer must ensure that the level of performance of each component of the system is adequate to meet the overall system performance requirement.

On the other hand, owners and operators usually consider operational issues of primary importance and generally have some strong opinions about what qualities are necessary to achieve the desired performance. The best way to develop basic performance-level standards is through discussions between the engineer and the owner's representatives. In these discussions, the engineer can explain the design from an operations standpoint and solicit the owner's views concerning operating and maintenance features of the system.

The engineer should act as a resource for the owner—to answer questions or concerns with up-to-date information about the advanced technological system being considered. In this way, the basic standard of acceptable system performance can be developed for the design. Our firm's approach is to prepare draft specification to incorporate those changes. We have found that the process works well because it prepares the owner for participating in the process of selecting a system from those that will eventually be proposed.

Maintenance and Expansion

Most advanced-technology energy-related systems include a substantial amount of equipment that has been designed and constructed by only one manufacturer. To repair, expand, or replace this equipment generally entails noncompetitive sole-source procurement. While this is not a major portion of the business of most manufacturers, some local representatives cannot resist the chance to take advantage of the owner when such sole-source purchasing is required.

By making unit pricing and extended-warranty offers a required part of the procurement process, the engineer and owner can eliminate these potential problems. They have an opportunity to discuss options for supporting, maintaining, or perhaps expanding the advanced-technology system before the system is purchased and for ensuring that the ongoing needs of the owner are addressed in the procurement process.

Note that if it is decided that unit pricing or extended-warranty offers are desired for more than several years past the date of completion, then potential vendors should be given wide latitude to tie their pricing to price indices such as the consumer's price index, interest rate, or other industry pricing indices. If maintenance of the system is expected to be substantial or if the owner intends substantial expansion of the system, then the evaluation of the warranty offer and/or unit-pricing schedule may be an important part of the evaluation process.

Owners have found the extended-warranty offer requirement to be very valuable in comparing competing systems. Because the offer is made in a competitive process, each extended-warranty offer usually represents the contractor's best estimate of the true annual costs of maintaining the system. Whether or not the owner intends to contract with the vendor to maintain the system after the one-year construction warranty expires, the vendor's estimate of annual warranty cost is very telling about its confidence in the integrity of the equipment to be supplied. A requirement that the extended-warranty agreement remains open until the expiration of the construction warranty period permits the owner a year of operating experience before deciding whether to contract emergency repair services or provide them internally.

Step 2: Procurement

Once the decision to procure a project's advanced-technology components competitively has been made, the owner and engineer must work together to ensure that a competitive selection process is developed to meet the owner's procurement rules. If the owner represents a private organization, there are likely to be rules and guidelines within the organization that may limit the flexibility of the procurement process. If the owner is a government agency, then laws may limit the procurement. The best rule to follow is to make no assumptions regarding which rules will actually apply.

The goal of the procurement process should be to break the advanced-technology system(s) out of the basic construction bid documents and to conduct a separate procurement, generally with the plan to reattach it to the construction contract once that procurement process is complete. The cash-allowance feature of construction contracts is an excellent vehicle for accomplishing a separate procurement. The use of cash allowances is widespread in construction contracts today.

The presence of the cash allowance alerts the general contractor to the fact that management resources will be required to coordinate this element of the construction effort, but the general contractor does not decide who the subcontractor will be. Once the procurement is completed, the actual subcontract amount is substituted for the allowance, and the general contractor assumes responsibility for managing the entire construction effort.

Figure 7-2 illustrates how the cash-allowance process permits the separation of advanced-technology system procurement from the general contract. There are other ways to accomplish advanced-technology installations, but if the system or component is an integral portion of the overall building operation, a cash allowance is recommended because it does not compromise the general contractor's control of or responsibility for the construction effort.

To ensure that an effective procurement process is established for

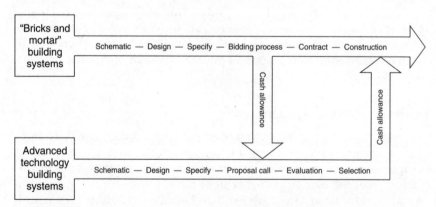

Figure 7-2. Diagram showing use of cash allowance to conduct separate procurement processes for advanced-technology building systems.

high-technology systems, I suggest the following guidelines be followed when the process is developed:

Make the Case

There is an important reason why you are likely to hear negative responses if you ask about procurement flexibility in large bureaucracies. People like to do things the easiest way possible. Conducting construction procurement with a traditional bidding process is usually the easiest way. However, if the issue is presented candidly, I have found that the problems associated with garnering support within the bureaucracy's purchasing establishment are generally greatly reduced.

To make the case, the engineer should be prepared to discuss the advantages of a proposal process realistically along with the problems of a strictly bidding procurement. Having as many of the differences expressed in bottom-line dollar figures is the best way I know to get the attention (and sympathies) of people in purchasing.

In making the case, provide a specific procurement plan that you believe will meet the owner's organizational rules. The most powerful support you can garner comes from patterning your plan after a similar process that has been recently employed by the organization with a successful outcome. For this, some research and investigating is necessary.

When you are working with government projects, consider contacting the offices of representatives who chair committees involved with energy to find laws that permit (and sometimes encourage) nontraditional procurement procedures. State-level energy departments are good resources when a state agency is a client. Whatever resources are available, use them to sell the idea of nontraditional procurement rather than ask about it.

The Evaluation Process

One of the reasons why organizations are leery of nontraditional procurement processes is the fear of conflict-of-interest situations. Include a well thought-out evaluation process to allay this fear. Form an evaluation committee that includes the building operating personnel and at least one person from the department in charge of purchasing for the project. This is a good way to eliminate worry that someone may put something over on an organization.

In addition, a description of the range of features that will be under scrutiny in the evaluation process and the financial considerations that accompany each feature is a valuable tool in making the case for nontraditional procurement. The more it is possible to put the evaluation process into a formula, the easier it will be to make the process fly. Of course, it is important to understand that many of the items that may distinguish one proposal from another cannot be listed before the proposals are received. Still, that does not prevent the engineer from establishing broad categories of possible differences and making an assignment of relative value to each category.

A Realistic Process

Every client organization has a personality complete with preconceptions and prejudices that are usually the result of its past experience—good and bad. The interpretation of these experiences has often led organizations to a particular way of doing things that can conflict with the engineer's vision of how to develop the procurement process most effectively.

The engineer must be sensitive to the organization's way of doing business. Only if harmony would compromise a satisfactory outcome should the engineer challenge established procedures, and if such challenges are necessary, they should be minimized so as not to cause the whole process to be discarded.

The cornerstone of successful procurement of advanced-technology systems involves breaking those components out of the traditional bidding process. Depending on the technological system to be purchased and the organizational constraints and views by which the owners' procurement procedures are governed, a successful procurement may be as simple as a prequalification of acceptable system configurations and components. There is no need to rock the organization's boat unduly if a very simple procurement process that is compatible with the organization's general business attitudes will suffice.

Step 3: Develop Specifications

When the process by which the procurement will take place has been determined, the engineer must translate the design of the advanced-technology system into a specification document that meets the requirements of the chosen process. The following basic rules should apply to the specification development process:

Design the system only once. Earlier, we discussed the need for the engineer to be as specific as possible in selecting the advanced-technology systems. This requires the engineer to be knowledgeable about the technology to be used and to provide a complete design of the application (with alternatives if necessary).

If the design is not complete, the contractor will have to provide a design as a part of the contract. The contractor's design will require full knowledge of the way in which the system must integrate into the other mechanical system components, but that information may never be made completely available to the contractor. Therefore, the second (contractor's) design almost always has defects that usually don't come to light until start-up, at which time they can cause enormous headaches for the owner and operators.

To eliminate this problem, the engineer should assume complete responsibility for each advanced-technology system that is a part of the overall mechanical design. If the engineer does not have the expertise to provide the design, then an independent outside source should be found to help with the design, specifications, and procurement.

It is important to remember that the design of the overall mechanical system is often the result of many months of information exchanged among the engineer, owner, and others in the design team. It is not reasonable to expect the system subcontractor to pick up all this information and assume responsibility for the design of the system.

Maintain competitive flexibility: It is important for the engineer to provide a complete design so that the potential contractors know exactly what will be expected of them. Also, specifications should not be so specific as to favor one contractor's equipment over another's without valid reasons.

For example, DDC system panels have varying point capacities among the various manufacturers. When specifying for a competitive DDC procurement, our firm tries to detail each point and its interface and to establish the level of control, communication, and operator interface required. But we do not specify a point connection list for each panel. If we were to do so, worthy systems might be eliminated simply because they incorporate lower point densities at each panel. Or systems with high panel point capacities might be placed at a competitive disadvantage by requiring a low point density. Unless there is a compelling reason to require a precise point-to-panel configuration, the process should allow each contractor to match his equipment to the points list most cost-effectively. Then the configuration becomes an item of the evaluation process.

Designers should try to develop specifications so that as many products as possible can be considered. Obviously, the more versatile the procurement process, the more flexible the specifications can be. But flexibility should not be confused with ambiguity. No matter how flexible the specifications are, they must very clearly describe what is required of the system or component.

Uncertainties

An owner once told me that in discussions with his engineer on a project, he questioned the engineer about the lack of a points list for the DDC system. The engineer replied, "Oh, we never include points lists in our specifications. This way the vendor must supply all points that are necessary to make the system work. We never have to worry about leaving any out."

That revealing statement should cause engineers and owners alike to pause and rethink the interests and duties in construction projects. We all know that there are some uncertainties in most building system designs. Those who are most aggressive in achieving high levels of energy efficiency, comfort, and air quality probably encompass somewhat larger uncertainties because many utilize technologies or strategies that have not been widely employed before. Presumably owners who encourage or permit higher performance designs have some understanding of the uncertainties involved.

However, the high-technology contractor is the least prepared to evaluate and accept responsibility for a project's uncertainties. The contractor has not been privy to all the discussions and decisions throughout the design process and very likely does not understand the complete mechanical system design. When the specifications are released, each contractor is (or should be) wholly engrossed in a good-faith effort to apply her or his product in the most suitable configuration possible.

Putting the uncertainties of the project on the contractor's back results in a costly redundant engineering effort. At times it seems that the use of ambiguous procurement procedures is aimed at enticing contractors to underestimate the scope of work and obtain lower prices. More frequently, however, I find ambiguous specifications are the product of ignorant designers.

Whatever the reason for a specification that does not clearly define the scope of work for an advanced-technology system or component, it is a risky procurement strategy because the project and the owner are almost always the real losers. The staff members of most contractors have been around long enough to make provisions for risk. Their bids or proposals respond to ambiguity with ambiguity. I have heard vendors state that if the engineer requires them to provide part A, which they

I. System design
   A. Does the design meet the functional requirements?
   B. Does the design ensure that the necessary system function will be achieved?
   C. Can the owner or user operate the system?
   D. Is the owner or user satisfied that the system design will be successfully operated?

II. Specifications
   A. Do the specifications invite competition?
   B. Do the specifications ensure that three or more vendors can provide competitive pricing?
   C. Are the specifications and supporting documents clear?
   D. Do the specifications describe exactly what is required of the contractor and others to complete the contract?
   E. Do the specifications provide flexibility?
   F. Do the specifications include provisions for voluntary options that permit each vendor to propose the most suitable configuration of its equipment?
   G. Do the specifications support ongoing operations?
   H. Are provisions for operating features desired by the owner included?
   I. Are unit pricing and extended-warranty offers required in each proposal?

III. Evaluation and selection
   A. Does the evaluation process require disclosure?
   B. Is each vendor required to disclose fully in her or his proposal what is to be provided?
   C. Can a reasonable selection be made?
   D. Does the evaluation team possess the expertise to evaluate the technical merits of each proposal?
   E. Is the evaluation process value-based?
   F. Does the evaluation process encourage selection based on overall value and not just lowest first cost?
   G. Is the evaluation process technically focused?
   H. Is the evaluation process designed to make a selection on technical merit and not on sales presentations or materials?

Figure 7-3. Checklist for successfully incorporating advanced technology systems into a building construction project.

believe was not required by the specifications, they intend to respond by downscaling or omitting parts B, C, or D, whose exact requirements are obscured by the ambiguous nature of the specifications.

What those conducting the process must consider is that as long as the requirements of the advanced-technology system are realistic, then the more complete the specifications are in describing the level of performance and other requirements of the system to be purchased, the more likely the suppliers will be to provide competitive proposals. The specifications must also provide sufficient latitude that several potential suppliers can offer cost-effective proposals. The engineer must stay up to date with each technology of the design and possess sufficient knowledge of the range of products available to ensure that the balance is maintained between desired features and opportunity for multiple vendors.

While a great many engineers still use simple bidding procedures to purchase controls and other advanced-technology building systems, the bidding process can result in costlier systems that fail to perform to the owner's expectations. Placing a project's high-technology systems and components in a special category and procuring them separately from the general contract comprise an approach that has proved useful to the owners and engineers who use it regularly.

Figure 7-3 illustrates the three basic steps that firms who regularly work with advanced-technology systems have found to be effective for successful high-technology system procurement. The questions are used as a checklist to ensure that the resulting system meets the demands of the project and will be procured effectively. Seeing that these questions are answered positively and that attention is paid to the special procurement process will result in improved performance and cost of each project's high-technology systems.

8
Operator-Interface Techniques

The key to effective high-performance DDC system operation is to keep it simple! Effective operation means the DDC system is available directly to all building mechanics or maintenance staff as a tool in the execution of their duties. To these people, the DDC system should be used to learn more about problems or complaints before they leave on a service call. Effective operation also means that the primary operator can easily interrogate the system and adjust the program whenever required. To promote effective operation, I usually recommend against elaborate reports and other more complicated operator-interface options, in favor of establishing fast and simple operator interfaces that can be used by all the operations personnel. In most successful installations, these simple operator interfaces and interrogation procedures are all that is used in day-to-day operations.

Point Key Names

All DDC systems today use alphanumeric key names to identify system points and variables. Selecting good key names is very important to daily operation, because when a point is interrogated, the key name in itself should be sufficient to precisely identify to the operator each point or variable. Yet the names should also be short and consistent, so that the operator can enter each point name correctly from memory. For this reason, I recommend that a naming convention be applied to all points and variables. I urge users to develop a four-level point key name convention from the basic syntax shown in Fig. 8-1.

Ask an operator at a successful DDC system installation. Chances are that each point name fully describes the point, its location, and function. A good logical points-naming convention is a great help to this understanding.

Real-Time Displays and Printouts

The most effective DDC information output is the real-time points display. The real-time display generally is thought of as a screen of points and values or as a graphics display with values (such as space temperatures) embedded in it. However, with the large numbers of points in full DDC systems, the use of color representations has become increasingly valuable. For example, the color green may be used to represent a space that is within its heating and cooling setpoint range. Red indicates an excessively warm room, and blue means a cool one. When the operator calls up the floor graphics or display, the conditions on the floor are immediately apparent—without the operator having to read through a lot of numbers.

However the display is constructed, it is extremely important that the system permit a rapid access to each display and that it be continually updated with current status or values for all the points in the group—usually at 5- to 30-s intervals. There must be some simple means of obtaining a printout of the display as well. For most DDC systems, this display yields information more quickly and with simpler procedures than other output forms. Real-time displays and graphics should be formatted in a logical and consistent order for easy operations.

Figure 8-2 shows a format for simple real-time displays that I have found easy to scan for a variety of potential problems. The control status of the main system is first. Beneath that are points indicating the operating features of the system. On the right is the status of all the various outputs that regulate the system when it is operating. At the bottom are points indicating the quality of the system operation. Such a display can be easier to use by maintenance staff than graphics because the access is quicker.

The operator can use this display format to quickly scan for a number of different concerns. The operator is able to look to the same areas of the screen or printout for similar types of data no matter what type of system is being reviewed. Figure 8-3 shows typical constant-

Level 1: Buildings

Develop a one- or two-character abbreviation for each building of the facility.
Note: This level can be omitted in one-building facilities.

Level 2: Systems

AHn	Air handling system ($n = 1, 2, 3$, etc.)
BLn	Boiler ($n = 1, 2, 3$, etc.)
CHn	Chiller ($n = 1, 2, 3$, etc.)
CTn	Cooling tower ($n = 1, 2, 3$, etc.)
CVn	Converter ($n = 1, 2, 3$, etc.)
RDn	Radiant heating system ($n = 1, 2, 3$, etc.)

Level 3: Subsystems

CC	Cooling coil	P	Pump
CD	Cold deck	PH	Preheat
E	Exhaust	RA	Return air
F	Fan	RH	Reheat
HC	Heating coil	RW	Return water
HD	Hot deck	S	Space
MA	Mixed air	SA	Supply (discharge) air
OA	Outside air	SW	Supply water

Level 4: Devices

D	Damper	SS	Start/stop (DO)
H	Relative humidity	T	Temperature
P	Pressure	V	Valve
SP	Setpoint	EC	Electric current (DI)
AF	Airflow		

Special

AVE	Average	MAX	Maximum
CDM	Cold-day mode	MIN	Minimum
DHT	Day high temperature	PHT	Projected high temperature
DHTT	Day high-temperature time	PLT	Projected low temperature
DLT	Day low temperature	OCC	Occupied mode
DLTT	Day low-temperature time	OBJ	Objective

Examples:
AH4SAT   = air handling system 4, supply air temperature
AH2MAXST = air handling system 2, maximum space temperature
A-STOBJ   = "A" building space-temperature objective

Figure 8-1. Recommended DDC system point names convention.

```
MAIN SYSTEM START/STOP COMMAND    SUBSYSTEM START/STOP COMMAND
MAIN SYSTEM STATUS                SUBSYSTEM STATUS

SUBSYSTEM TEMPERATURE            SUBSYSTEM ANALOG OUTPUT 1
SUBSYSTEM TEMPERATURE            SUBSYSTEM ANALOG OUTPUT 2

SUBSYSTEM SETPOINT
SUBSYSTEM SETPOINT

SYSTEM OUTPUT TEMPERATURE OR STATUS
SYSTEM OUTPUT TEMPERATURE OR STATUS
SYSTEM OUTPUT TEMPERATURE OR STATUS
```

Figure 8-2. Real-time display layout format.

```
AH1SAFSS    ON·        A         AH1CCP     OFF      A
AH1SAF      ON·        A         AH1HCP     OFF      A
AH1RAF      ON         A

AH1RAT      74.2 DEGF A          AH1MAD     48.6 %   A
AH1MAT      68.2 DEGF A          AH1CCV      0.0 %   A
AH1SAT      69.4 DEGF A          AH1HCV      0.0 %·  A

AH1SASP     69.1 DEGF A

AH1AVEST    72.2 DEGF A
AH1MAXST    73.5 DEGF A
AH1MINST    71.8 DEGF A
```

(a)

Figure 8-3. Real-time display layouts for typical fan and chiller systems.

```
CH1SS      ON·       A        CHSWP      ON·       A
CH1        ON·       A        CTSWP      ON·       A

CH1SWT     46.4 DEGF A        CH1DMD     62.6 %    A
CH1RWT     52.2 DEGF A

CT1SWT     84.6 DEGF A        CTF1       ON        A
CT1RWT     89.7 DEGF A        CTF2       OFF       A

CH1SWSP    46.2 DEGF A
CT1SWSP    85.0 DEGF M

AH1CCV     88.0 %    A
AH2CCV     72.1 %    A
AH3CCV     60.2 %    A
```

(b)

Note: CH1DMD is the DDC output to the chiller that regulates the demand setting of the chiller which is used in this instance to maintain the chilled water setpoint, and also to sheld electrical demand under certain conditions.

A—represents points under automatic control; M—for points under manual control is this DDC system display convention.

Figure 8-3. (*Continued*) Real-time display layouts for typical fan and chiller systems.

volume air system and chiller displays formatted according to the display convention of Fig. 8-2 and the points key name convention of Fig. 8-1. These displays are quite easily read even by a maintenance mechanic who does not regularly use the DDC system. Rather than having to sort out all the points before understanding the display, such secondary operators can very quickly identify the particular item(s) of concern. Note also that the example displays show whether the various points are in automatic operation (controlled by the control program) or manual operation (overridden by the operator). It is a good idea to fully understand the display and printout features of a DDC system before you purchase it, because the display is typically the most common operator-interface technique employed in successful systems. At a minimum, the real-time display should permit:

1. User-selected organization of points that allows at least 40 points in a single display or graphic and permits each point to appear in more than one display or graphic

2. An automatic value/status update frequency for all displayed points of 30 s or less

3. A quick and simple means for printout of each display or graphics screen

4. All displays to be accessible by secondary terminals, including portable terminals connected via phone lines

5. The control condition of each point to be noted on the display (for example, manual or auto control)

At successful DDC system installations, I find operators use displays to troubleshoot potential problems quickly and effectively. They can scan each display quickly for the pertinent information and go to another to resolve a problem much more quickly than operators at less successful sites. The ability to interrogate a system quickly, make adjustments, and have an override condition noted on any display of the point is an important organizational attribute of a successful DDC installation.

Some users have established point key name standards and display formats different from those of Figs. 8-1 and 8-2, usually because of system limitations on allowable key name characters or length. The exact format of key names and real-time point displays is usually less important than the consistency of their application.

Modem and Portable Terminal

One of the most substantial automatic benefits of DDC systems is the ability to call up the system via telephone modem and to obtain displays, make adjustments, or install temporary program changes. This feature provides a number of opportunities to the user. It allows an operator at one site to monitor and adjust the operation of one or more other DDC system sites without visiting the site or even changing computer terminals. This capacity is particularly useful when a user's facilities consist of several small buildings that cannot justify individual operators at each. It is also useful for facilities whose hours of operation extend beyond those of the maintenance staff. If a problem or question develops, the operator can connect to the system from home with a portable terminal, make required adjustments, and very often eliminate the need for a trip to the site.

Users who now have DDC systems or are contemplating the purchase of DDC systems in the future should develop their operating plans with a high reliance on the use of modem operation for system support. The more successful high-performance DDC system installations employ operators who, with the use of the modem connection, provide a better level of oversight to the facility than has ever been attained previously and who are far better able to ensure that small problems do not become big issues before they are detected.

Trend Logs

A trend log is often the only way that certain types of problems can be properly identified. For example, a freeze thermostat that trips occasionally in cold weather may be caused by an unstable mixed-air damper control loop, an error in the mixed-air temperature setpoint calculation, faulty mixed-air damper linkage, or a failure in the freeze stat itself. If the problem occurs only occasionally, a trend log of certain points on the air system is the most direct path to its solution. Usually, such a problem can be entirely solved by reviewing the trend log of the points affected during the freeze stat trip.

Trend logs typically are intended to be set up to troubleshoot just such a problem. The operator chooses the points, selects the interval between records and the total number of records, and starts the program going. Almost all systems permit the use of a rollover log which stores records up to the operator-selected number and then overwrites the oldest record with each new record. Thus only the most recent records are maintained. If the freeze stat trips once a week, the operator need not include a full week's number of records, but simply stops the trend log as soon as she or he notices the stat has tripped. The most recent records will include the conditions immediately up to the freeze stat trip.

Operators I have observed at successful DDC system installations also run some general trend logs continuously. These trend logs may include several days' values of all the space temperatures every 30 min or the status of boilers, chillers, or other major energy systems. The points vary from system to system, and usually they include areas of operation that are of the most concern to the operator. High-performance DDC system installations may wish to employ continuous trends of certain system points which are periodically dumped to magnetic media. This allows the operator to call a trend for any group of points over any period without prior setup. However, this feature usually adds expense to the system and needs itself to be supported in order to operate effectively. Continuous trending is a useful feature when the operator does not have the time to set up and monitor individual trend logs for the number of problems that the operator is requested to investigate.

To be effective, the DDC system trend log capacity must be extensive. It is helpful to have the capacity to keep trend logs of all system points at once. At a minimum, one-third to one-half of the points must be capable of being tracked simultaneously. The automatic rollover feature described earlier is a must, and the display should permit either a tabular or a graphics representation of the values or status. The trend log must be easy for an operator to set up and view at any time.

9

High-Performance DDC Documentation

Previous chapters have described various aspects of implementing high-performance DDC strategies. This chapter discusses the role of documentation, an essential item of support to ensure the long-term integrity of any high-performance DDC system installation.

For any high-performance DDC system to be successful, it must be adequately supported, which requires good documentation. Here we discuss the role of documentation and explain what users have found to be the best means of attaining the level of documentation required to adequately support a high-performance DDC application.

The Role of Documentation

Adequate documentation is one of the most important requirements to achieving long-term success with high-performance DDC systems. But developing good documentation is very often overlooked by system designers. Most system specifications today simply state the information and data that must be included in the documentation. But the quantity of documentation has very little to do with the quality. I have seen DDC systems that required entire bookshelves of manuals and drawings. Documentation this voluminous rarely provides adequate support to those with direct operational responsibility. On the other hand, simply paring down the contents of the documentation does not necessarily improve it or lead to better support. The reason why documentation is so important to the success of DDC systems is that it must *organize* the elements of system operation so that any need is easily accessed by an operator who has much more to do than to become a self-taught expert on the intricacies of the DDC system.

Two basic categories of documentation are required to support a DDC system installation:

1. System documentation
2. Application documentation

System documentation consists of the manuals developed by the manufacturer to explain and support the operation of its system. Operator's, maintenance, and reference manuals are typical system documentation. These manuals are essentially the same for all installations of a particular system. Application documentation includes reference material pertaining to the specific installation. Layouts and schematics, check-out and database materials, and a description of the high-performance control programs should all be included in the application documentation.

System Documentation

A number of different approaches to providing system documentation have been taken by DDC system manufacturers. Some have responded to user demands for more complete documentation by providing user manuals in a reference format that results in substantial volumes of material. Others have tried to develop manuals that describe the system functions in layers. The problem common to these documentation approaches is that the operator usually does not have a quick reference for immediate problems or typical questions that need to be answered to use the system effectively. Why? Because most system documentation is not prepared with an operator in mind. More often, it is meant as a reference for the installation or service technicians.

As an example, consider an operator who wishes to find the procedure to produce a report that is not used regularly in her work, say, a combined graph of several trend logs. Assuming the system has the capacity, she probably will not find the procedure in one simple write-up of the system documentation. Instead, she must hunt through an index of commands, but does she look under *trend logs* or *graphics?* Usually, multiple commands must be executed to develop such a document, and the various required command sequences are described in several different areas of the documentation. In such a situation, the operator often gives up the search, and the system feature is never employed.

Obtaining Good System Documentation

A manufacturer is unlikely to develop a complete set of system documentation to satisfy a single user. Thus, the most prudent approach for a user wishing to acquire effective system documentation is to see that supplementary information is developed as needed to ensure the final documentation meets all needs. This process begins by reviewing the operation and documentation of the proposed DDC system before it is purchased. By visiting a site (without the manufacturer's representative present) and working a short time with the operator, one can quickly assemble a list of items that need to be addressed in addition to the standard system documentation. These items can be discussed, and the responsibility for completing the required documentation can be assigned. Depending on the installation, the responsibility may be given to the contractor, the consultant, or even the user.

For most DDC systems, some supplemental system documentation is required to describe certain operations that the operator must perform which are not described adequately in the manufacturer-supplied system documentation. System documentation should be supplemented as required to be certain to include a reasonable description of how to accomplish all the following items:

Basics: Signing on and off the system. Rebooting and database backup procedures, and what to do to break a system or panel lockup.

Database: How to create and delete system points and variables, and how to print out any of the database listings.

Displays and graphics: How to set up or change real-time displays and graphics; how to print them out.

Trend logs: How to set up or change trend logs; what rules apply to number of points, trend frequency, triggering, etc. How to display and print trend logs.

Manual control: How to achieve manual control (override program control) of each type of system input/output point or variable.

Alarms: How to set up and change the parameters for analog and digital alarms, including a list of automatic system alarms complete with a description of what each means.

Programs: How to write and edit programs in the system's control language, including a description of all language features and formatting requirements, How to write and edit time schedules and any other system programs or features to be employed in the application.

Secondary operator's guide: A very simple guide for backup operators so that one who only occasionally operates the system can fill in for the operator when necessary.

Usually, most of the above items are adequately documented in the manufacturer's manuals. For those items in the above list that are not adequately documented, the user must see that adequate supplementary documentation is provided. The operator must be able to perform any of the above features when required without difficulty.

Application Documentation

While the system documentation is important to adequately support a high-performance DDC system, developing good applications documentation is even more crucial to its long-term success. The DDC system manufacturer can usually be persuaded to provide additional information about system characteristics if operating support becomes a problem, but obtaining after-the-fact information about the application documentation can be very difficult, especially after the operator has begun to make modifications to the programs or database.

For this reason, it is recommended that the following application documentation be included in a high-performance DDC system installation:

System Configuration Schematic

The configuration schematic consists of one or more drawings that show the location and interconnection of all DDC system components except the end devices. The connection of all operator-interface equipment, master and slave panels, unitary controllers, etc., must be shown in schematic form on the drawing, appropriately labeled, and located. The drawing(s) must also show the location of major raceways and termination panels.

This schematic provides an important overview to what otherwise may seem to the maintenance staff to be an endless web of wires and panels. I recommend that contractors be required to provide this schematic with their proposal. It can then be refined and updated as needed to become part of the final documentation.

Equipment Description

The second aspect of the applications documentation that I recommend is a specification catalog of all equipment employed in the system. This catalog includes all the equipment shown in the schematic and all end-device hardware also. At a minimum the specifications for each unit must include manufacturer and model number, input requirements, accuracy and calibration data, capacity, environment limitations, parts lists, etc. As with the schematic, I recommend suppliers be required to provide this information in their original proposal and that it be adjusted as necessary to be a part of the final application documentation.

Labeling, Points Checkout, and Calibration Manual

One of the aspects of DDC system installation that seems most often to be neglected is calibration and testing of the integrity of the system. Too often we are told that devices are "factory-calibrated" and require no further adjustment. This is almost always bunk. Contractors can sometimes get away without completely testing and calibrating lower-performance systems (such was often the case in the days of pneumatics). But a DDC system deserves to be properly tested and calibrated to ensure it is providing its full capabilities for accuracy and function. PID control has brought about the bad practice of employing large proportional bands on analog outputs that may then take too long to react to changes in setpoint. For systems operating high-performance control strategies, good testing and calibration are a necessity because the errors from poor calibration and tuning may appear to be programming problems and thus mask their solution.

The system checkout process is aimed at uncovering and correcting operating, calibration, and loop-tuning problems prior to system start-up and to the introduction of the high-performance control program. To ensure that points are properly calibrated and perform as expected, I recommend that the installer and user each provide one person to conduct and document a checkout of all system points. For this procedure we employ a book that has a separate page for each point. The completed book becomes part of the application documentation and is used by the operator as a record of further calibration, maintenance, or replacement of each system point and its device. This procedure ensures that this important step in EMS commissioning is not overlooked.

Points Database Printout

To prevent uncertainties due to changes that are not documented, I recommend that all applications documentation include a complete printout of the database at the time of turnover. This printout will include the value of all parameters for each system point, configuration data on panels, operator-interface devices, etc. This list must be separate for each panel so that if points are added or other changes made to one of the panels, the operator can update only that portion of the database printout.

Control Language Program Listings

As we have seen in previous chapters, high-performance control programs rely on the specific program listings for the major portion of the documentation. How effective these listings are for use by the operator in checking the operation of a system point depends on how readable the given control language is. However, even when a control language does not provide in itself an effective explanation of the system operation, it is still the final determiner of how the system operates, and the applications documentation must include complete listings of the control language programs, time schedules, and any other routines employed in operating the high-performance control strategies.

Description of Programs

How much description is required to supplement the control language printouts in order to document the high-performance control strategies depends on how effective the control language is in providing that documentation itself. Relying on the program itself as documentation has the substantial benefit of self-documenting any changes made in the program. This self-documentation capability of DDC systems is a goal that I encourage manufacturers to strive for. In fact, in some systems with good languages, highly motivated operators have become sufficiently conversant in control language syntax to read the control sequence directly from a program printout. It is not, however, a good idea to plan on this happening for your application. Instead I recommend a two-tier program write-up to support the program listings. The first level consists of a logic-flow outline. I recommend against using flow diagrams for this or any other element of the applications documentation. Flow diagrams are too difficult for the operator to update when changes are made, so it never gets done. Write-ups are kept in text files, usually on the DDC system front end, and are more likely to be kept current.

The second element of program listing support documentation is tailored to fit the deficiencies of the particular control language employed. This element acts as a bridge between the logic flow outline and the listing. For several current control languages, this bridge is not required, but for most it still is. This write-up does two things. It directs the reader to the exact line or lines of the listing that accomplish each of the steps of the logic-flow outline, and it explains mathematics or logic operations that are not clear from reading the listing itself.

All the program descriptions should be in modules that exactly duplicate the high-performance control software modules. A means of cross-referencing can be found so that it is easy to determine the write-up and the program for each system point. As stated early in this book, all control programs are output-oriented so that the mathematics and logic which calculate or control each system point or variable are located in only one module. Therefore, if a system point is not behaving as it should, the operator must inspect only one module and interpret that portion of the program to begin troubleshooting the problem.

Points to Remember

The development of documentation for a high-performance DDC system should begin at the same time as the decision is made to implement the system. The original specifications contain the beginning features of the applications documentation, such as the points list and points names, and establish a format so that the winning proposal will contain the beginning elements of the system documentation. As the implementation of the DDC system proceeds, the refinement of this documentation and the introduction of additional documentation can be done smoothly as long as the responsibility for each element of the documentation is well established and the owner and consultant take time to review the material with the operations staff.

Although many DDC system installations today suffer from inadequate documentation, there is no reason why the system with which you are associated must suffer from this problem. The more complex strategies of high-performance control do not require more documentation than any other system, but the increased interaction of the points can magnify the shortcomings of an inadequately documented system.

Appendix

Definitions

Control Language   The set of characters, conventions, and rules that is used for processing information and initiating automatic digital and analog control of mechanical and electrical systems.

EMS   In this document an EMS (energy management system) is the hardware equipment and software that permit physical point information to be gathered, control sequences to be processed, and an operator interface. A system is limited to those components that comprise or can access a single database. Systems that employ multiple panels sharing only limited information are considered separate systems.

Point Names   Assigned alphanumeric representations for each system point or variable that are used to represent the point for all program and database needs.

System Points   All physical points that are connected to the energy management system are called *system points*. These include inputs that sense temperatures or statuses and outputs that control motors or valves.

Variables   Points that do not exist in a physical sense, but are available for use by various elements of the control program are called *variables*. Setpoints and run times are simple examples of variables.

Operator's Control Language

General Requirements—Points, Description, and Access

Operator Overrides.   A simple uniform means for changing point operation from automatic to manual control for all system points and variables must be provided to enable easy override by the operator in case of a problem. The override shall permit the operator to assign any value for the point within its defined limits. Each point under manual control shall include an additional special notation every time the value/status of the point is displayed.

Digital Points. A digital (binary) point must retain a value (1 for on and 0 for off) in addition to its status so that the point can be used directly in mathematics functions. Digital points 1, TRUE, and ON are interchangeable equivalents, and 0, FALSE, and OFF are equivalents (the digital-point definition must allow a switch to define the deactivated state as ON if desired).

Analog Points. Analog inputs and outputs must be provided with appropriate scaling factors such that the operator can assign one or more factors for different ranges of values. For example, nonlinear input and output devices may require several different factors and offsets to provide required accuracy over their entire operating range. An analog output to a pneumatic device might utilize the first 5 percent of the output range to drive the output from 0 to 8 lb/in^2, the next 90 percent to drive the output from 8 to 13 lb/in^2, and the last 5 percent to drive the output from 13 to 20 lb/in^2.

Variables. There must be no fixed limit on the number of nonphysical variables that can be defined. Variables shall have the same capacities for display, manual control, program access, command, calculation, and decisions as system points. Variables used as binary point must retain the value 1 for TRUE or ON and 0 for FALSE or OFF in addition to user-defined descriptors. Variables must be able to be set to themselves (example: VAR1 = VAR1 + 1).

System Variables. In addition to system points and variables, the system must automatically provide points that can be used by programs as follows:

Time of day: A system variable TIME that automatically provides the time in standard format (HR:MN) or decimal (HR.HR) depending on the requirement. Alternately, two forms of time of day should be provided. TIME, for use in logic expressions, provides time in hours and minutes (HR:MN) for comparing to present time, and HOUR, a decimal value in hours and hundredths (HR.HR), is used in calculating time-based factors etc. Both forms must be based on a 24-h clock, and automatically adjust for the Daylight Savings Time switch.

Day of Week: WDAY, the day-of-week system variable providing the numbers (1 to 7) for the days of the week, switching every night at midnight.

Month of year (1 to 12): MONTH

Day of month (1 to 31): MDAY

Day of year (1 to 366): DAY

Date (01/JAN/00 to 31/DEC/99): DATE

HOLIDAY, set by annual schedule, sets WDAY to 8 or 9

Time of morning sunrise (both HR.HR and HR:MN): SUNRISE

Time of evening sunset (both HR.HR and HR:MN): SUNSET. *Note:* Both SUNRISE and SUNSET require localization factors for correct operation.

Point Names. Point names should be any combination of alphanumeric characters up to at least 12 in number. Once a point name is assigned to a system point or variable, it must be all that is needed to access that point for any database or program function. The system must provide an easy means to allow the user to change a point name such that the point will automatically be referenced everywhere in the system by the new name.

Program Size. The OCL program space must be sufficient to support the following minimums:

1. 20 statements of any kind for each possible DO or AO that relies on that panel for logic
2. 10 statements of any kind for each possible AI or DI that relies on that panel for logic
3. A minimum of 200 statements of any kind at each panel or program resident location

These minimums are based on typical statements, each containing four points, variables, or constants and three operators.

Program Configuration Options. Each panel or program location must have the capacity of executing multiple programs. Numbers of possible programs should not be fixed, but allow for at least one program per possible system-output point. The size allowance of each program should be up to the maximum capacity for all programs in the panel as outlined above. Each program must have the capacity to be activated/deactivated, edited, or created/deleted without affecting the operation of other programs.

In configurations where several levels of intelligent logic and mathematical control exist, the control language must exist at all those levels. Rules of language use must remain constant throughout the control levels, although some specific features may be absent at certain levels. However, provision must be made to ensure accessibility to any system point or variable by point name at any level of control.

Preprogrammed Routines.   Any preprogrammed routines except special function operators must be written in OCL format to permit easy adjustments for special conditions or to suit particular mechanical or electrical systems. Preprogrammed control routines such as optimum start/stop programs must be provided with code and documentation so required changes can be easily and effectively made.

Program Readability.   The purpose of the language is to provide functional and understandable programs such that operators with minimal computer or language training can read, understand, and troubleshoot complex control sequences. It is therefore essential that the language be as readable as possible. For this, the following features are required:

1. *Complex statements:* Language must permit statements constructed of multiple variables and operators without a fixed limit. Also required is a means of line continuation that allows the operator to set line breaks.

2. *Mixed mathematics and logic in statements:* Language must allow mixing of logic and mathematics in statements and allow a mathematics or logic expression in place of a point name or variable.

3. *No restriction on order of operators in statements:* Language must allow user to construct statements in any functional order of variables, constants, and operators.

4. *Use of point names and automatic attribute selection:* Language must employ the point names for all points and variables. Program automatically selects desired attribute of point. For example, the statement that begins IF FAN ON automatically engages the system to consider the status of that point. Starting the statement IF FAN ALARM engages the system to consider the alarm attribute of the point.

5. *Minimum parentheses requirement:* At least five levels of parentheses must be available without restrictions to force the order of evaluation of mathematics and logic statements, but parentheses should not be mandatory in ordinary mathematics or logic statements. (*Note:* The use of parentheses is mandatory for special operators.)

6. *Comments:* Comments must be capable of being embedded anywhere in programs (including lines with expressions) through the use of special character(s) to start and stop comments.

7. *Floating-point mathematics and integers:* Language must employ floating-point mathematics, but also permit the use of integers with automatic recognition of integers as floating-point equivalence when required (5 = 5.0 in mathematics function).

8. *Line identification:* The use of line numbers in the program is discouraged as a detriment to readability. However, employing numbers in printouts and listings to identify specific lines for errors is encouraged. Automatic indents or other special characters to show a continued line or statements within BEGIN/END groups are also encouraged.

9. *Program printouts:* System must include a fast and easy method of printing programs and comments, single programs or in groups.

OCL Control Functions.   All control decisions and calculations must be executed through OCL. The use of separate routines and enhancements such as interlock programs, special calculated points, or database enhancements to meet the requirements of the OCL function is not acceptable. While many current systems can provide roughly equivalent function by linking these special functions together, such procedure results in programs that are difficult to understand, document, or use by the operator to make control adjustments.

Program Control

Timing Functions.   An operator shall be provided to permit time-based program sequencing in a readable fashion. This mechanism shall operate segments of a single program at precise timed intervals. The recommended operators are:

DOEVERY nn M(S)(H)

ENDDO

where the lines between the DOEVERY and ENDDO statements are executed every nn minutes (M), seconds (S), or hours (H). The term "nn" can be either a constant or a variable.
Example:

```
DOEVERY 10 M
    Statement 1
    Statement 2
ENDDO
```

Statements 1 and 2 would be executed on the first scan of the program and once every 10 min thereafter.

Timers. Variables acting as timekeepers must be available for user-constructed timing functions. Timers can be points that are specifically available for the purpose or as an option for variables such that unless they are reset to a specific value by a manual or program command, they time in hours and hundredths of hours, or minutes and hundredths of minutes, to at least 9999 h (min).

Branch Commands. Branch commands are not encouraged for readable programs, but they are occasionally necessary in most applications. The language must support GOTO, GOSUB/RETURN, CALL, and EXIT as follows:

```
GOTO (line label)
```

GOTO causes the program execution to jump immediately to the line label indicated. GOTO statements should be restricted to jumps forward in the program.

```
GOSUB (subroutine name)
```

GOSUB causes the program execution to jump immediately to the subroutine indicated. Program execution returns to the statement following the GOSUB statement as soon as a RETURN statement is encountered in the subroutine.

```
CALL (program name)
```

CALL causes the current program execution to halt and the program indicated to be executed. When the last line of the called program has been executed, execution returns to the statement following the CALL statement.

```
EXIT LOOP
```

EXIT LOOP causes the execution of the program to move immediately to the statement following the next END or ENDO in the program. If none follow, then it has the same effect as EXIT PROGRAM.

```
EXIT PROGRAM
```

EXIT PROGRAM causes an immediate end to the program execution. It produces the same results as reaching the last line of the program. Program execution is halted and restarted only when conditions for the program execution are next met.

Mathematical Expressions

General Rules

1. Essential to the OCL language is the evaluation and programming of complex mathematics expressions in a straightforward manner consistent with standard mathematical syntax. The language must allow expressions to be written free-form in standard mathematical notation.
2. Full 32-bit floating-point arithmetic must be supported and transparent to the language. Integers must be automatically converted such that they can be mixed with decimals in calculations.

Example:

```
MASP = 99.5 - 2 * AVEST - OAT / 5
```

3. Mathematical operators and functions must be allowed to mix with digital variables/points. Digital variables/points will equate to 1 for true/on and 10 for false/off in any mathematical expression.

Example:

```
TBSP = 70 + 5 * CHILLER
```

where CHILLER is the digital status point in indicating chiller operation.
4. Mathematical operators and functions must be allowed to be included in conditional expressions (see also Conditional Expression section).

Example:

```
IF MIN(ST1,ST2) < (NLLTEMP + LOWTSRT * 2) THEN LOWSTRT = ON
```

Standard Mathematics Operators

Exponentiation (^, **)

Negation (−)

Multiplication and division (*, /)

Modulo arithmetic (MOD)

Addition and subtraction (+, −)

Special Mathematical Functions

AVE( ) returns average value of group of variables or numeric expressions.

MAX( ) returns maximum value of group of variables or numeric expressions.

MIN( ) returns minimum value of group of variables or numeric expressions.

ABS( ) returns absolute value of variable or numeric expression.

LMT( ) is limit variable to set range between variables or numeric expressions representing minimum and maximum allowable values.

RND( ) rounds value of variable or numeric expression to nearest integer value.

TRN( ) converts decimal to integer by dropping decimal portion of number.

TON( ) means time point has been on (true)

TOF( ) means time point has been off (false).

SIN( ) returns sine of the angle variable or numeric expression.

COS( ) returns cosine of the angle variable or numeric expression.

TAN( ) returns tangent of the angel variable or numeric expression.

LN( ) returns the natural logarithm of a variable or numeric expression.

MOD( ) returns the remainder of an integer division.

INT( ) returns the largest integer less than or equal to the variable or numeric expression.

TBL( ) returns corresponding value from user-defined lookup table with at least 10 steps.

These functions would return values for use in mathematical or conditional expressions.
Example:

```
A = C + D + MIN(E,F)
```

Mathematical expressions must be permitted in place of variables in all mathematical functions.
Example:

```
B = MAX(C,A+B)
```

or a combination

```
D = MIN(E,AVE(F,G,H+I)) + J
```

Order of Precedence. Mathematics statements must be evaluated in a fixed order of precedence unless changed by parentheses; mathematical statements must always be evaluated before relational and logic operators in conditional statements. Statements must be evaluated in the following order:

1. Arithmetic operations
   a. Exponentiation ($\wedge$, **)
   b. Negation ($-$)
   c. Multiplication and division (*, /)
   d. Modulo arithmetic (MOD)
   e. Addition and subtraction (+, $-$)

2. Relational operations
   a. $=$, $<$, $>$, $<=$, $>=$, $<>$

3. Logic operators
   a. AND,OR,NOT

Operators at equal levels in a statement are evaluated from left to right.
Example:

```
VALUE = 3 + 6 / 12 * 3 - 2    is evaluated:
   a. 6 / 12 ( = 0.5)
   b. 0.5 * 3 ( = 1.5)
   c. 3 + 1.5 ( = 4.5)
   d. 4.5 - 2 ( = 2.5)
VALUE = 2.5
```

Parentheses in Arithmetic Operations.   Parentheses change the order in which arithmetic operations are performed. Operations within parentheses are performed first. Inside parentheses, the order of precedence is maintained.

Conditional Expressions

General Rules

1. No restrictions can be placed on statements executed by conditional expressions that do not ally to statements by themselves.

2. No additional restriction can be placed on the use of mathematics expressions when used in conditional statements.

3. Mixed mathematics and logic are allowed in IF statements.

Example:

```
IF FAN1 = ON AND (OAT + 5) > RAT THEN ...
```

4. In complex logic statements, mathematical operators must be evaluated first, then relational operators, and finally logic operators. Operators of the same level must be evaluated in straight left-to-right order unless changed by parentheses. See Sec. 2.3 D.

Logical Operators

ON	`IF FAN = ON THEN ...`
	`IF FAN THEN ...`
TRUE	`IF FLAG1 = TRUE THEN ...`
	`IF FLAG1 THEN ...`
OFF	`IF PUMP2 = OFF THEN ...`
FALSE	`IF FACTOR2 = FALSE THEN ...`

Note: TRUE is equivalent to and interchangeable with ON, and FALSE is equivalent to and interchangeable with OFF. TRUE or ON is implied when not included.

AND	`IF FAN = ON AND FLAG1 = TRUE THEN ...`
OR	`IF FAN = OFF OR FLAG1 = TRUE THEN ...`
NOT	`IF NOT FLAG1 THEN ...`

BETWEEN   `IF TIME BETWEEN 3:00, 4:00 THEN …`

`IF OAT BETWEEN 50, 60 THEN …`

Note: When BETWEEN statements are used with the system variable TIME, they must automatically adjust for the time change at midnight so the statement

`IF TIME BETWEEN 17:00, 08:00 THEN …`

is true between 5:00 p.m. and 8:00 a.m. the next morning.

ONTIME >   `IF FAN ONTIME > nn M(S)(H) THEN …`

This expression is true only when FAN is turned on and has been on for more than nn min, where nn is either a constant or a variable and the term following nn is M for minutes, S for seconds, or H for hours.

ONTIME <   `IF PUMP3 ONTIME < TIME1 M THEN …`

This expression is true only when PUMP3 is on, but has been on for less than the variable TIME1 minutes.

OFFTIME <   `IF FLAG1 OFFTIME < 30 S THEN …`

This expression is true only if variable FLAG1 has been off for less than 30 s.

OFFTIME >   `IF PUMP4 OFFTIME > TIME2 H THEN …`

This expression is true only if PUMP4 is off and has been off for more than TIME2 hours.

Mathematical Operators

<   `IF OAT < 55 THEN …`

>   `IF SPACETEMP > 70.5 THEN …`

=   `IF VARIABLE1 = 8 THEN …`

Note: The = sign can be used as both a mathematics and logic operator. In both instances it denotes equivalent.

Conditional mathematics operators can be paired to apply further conditions:

`IF A >= B`   If A is greater than or equal to B …

`IF A <= B`   If A is less than or equal to B …

`IF A <> B`   If A is not equal to B ...

Nesting Conditional Expressions.   At least five nesting levels of IF...THEN must be allowed. In nested IF...THEN statements, ELSE always refers to last IF THEN as follows:

```
IF expression1
THEN
IF expression2
THEN statement3
ELSE statement4
ELSE statement5
```

In the above expression, only if expression1 is true will expression2 be evaluated. If expression2 is true, then statement3 will be executed; if expression2 is false, then statement4 will be executed. If expression1 is not true, expression2 and statement3 and statement4 will be ignored and statement5 will be executed.

Multiple Statements—BEGIN/END.   To avoid the need for jump statements or subroutines, provisions must be made to allow multiple statements to be executed conditionally within the IF THEN ELSE format. When BEGIN and END are used following THEN or ELSE, all statements between will be executed as a single statement:

```
IF expression1
THEN
    BEGIN
    statement2
    statement3
    END
ELSE
    BEGIN
    statement4
    statement5
    END
```

Statement2 and statement3 will be executed only if expression1 is true. Statement4 and statement5 will be executed only if expression1 is false.

The program must note on each line by automatic indents when lines are part of a BEGIN/END group. This indication must show in the printouts of the program and whenever the program is displayed on the screen. A suitable syntax error signal must alert the operator if the numbers of BEGINs and ENDs do not match.

Special Conditional Statements

```
IFONCE FLAG1 = TRUE ...
```

This statement is true only the first time it is scanned after FLAG1 changes from false to true. FLAG1 must return to false and back to true for the statement to become true again.

```
IFONCE FAN1 = ON
```

This statement is true only for the first scan of the expression after FAN1 turns on.

Commands

Digital Commands. START and STOP are used to command digital points to their activated or normal states. Commands should be issued only when the program in which the command lines reside has completed execution.

START(FAN1) commands point named FAN1 to its activated state.

The language shall employ the principle of mathematical equivalent such that the following statements are equivalent:

```
(1)  START (FAN1)     STOP(VARIABLE1)
(2)  FAN1 = 1         VARIABLE1 = 0
(3)  FAN1 = ON        VARIABLE1 = OFF
```

Expressions on lines 2 and 3 are not recommended for use in turning system points on and off, but are sometimes useful in making variable expressions more understandable.

The language shall allow multiple points to be commanded in a single statement:

```
START(FAN1,FAN2,FAN3,...)
```

Note: For fail-safe operation, digital points shall be configured such that a command to the activated state may result in either the output contact closure or the output contact opening depending on the user-defined configuration.

Analog Commands. Analog output points are commanded to outputs using mathematics expressions. Actual output values must be user-set by scaling factors in the point database (see Sec. 2.1C).

```
VALVE1 = 100
```

This command sends the output to VALVE1 represented by 100 (100 percent if so set by scaling factor).

```
VALVE2 = 50
```

This command sends the output to VALVE2 represented by 50.

Note: Analog output points shall be configured to allow scaling and inverse factors so that a system with both normally open and normally closed valves with differing operating ranges can be scaled to all read and be commanded from 0 to 100 percent open at the operator's terminal.

Schedules

General Requirements.   A schedule system is a mixture of operating system and program language. The operating system must provide the scheduling mechanism. That mechanism must be easy to understand and easy to operate. Equally important, the schedule system must interface with the programming language such that time elements are readily accessible. For example, the decimal value of the start and stop times must be available to the OCL program so that comparisons and calculations based on their values can be made.

The schedule system must employ full-screen display and editing. Each schedule requires a unique descriptor of 30 characters to identify the schedule on screen. A full week, including holiday and special days, must be displayed on one screen. Simple procedures must allow a schedule to be created, edited, or deleted with ease.

Annual Schedules.   An annual schedule calendar mechanism must be provided to schedule holidays and special days a minimum of 1 year in advance. A full-screen editor in calendar format must be provided to allow speedy selection and review of holidays and special days.

Holidays, when in effect, force all schedules from their standard day of week to the holiday times of each individual schedule. The program language must have access to this holiday status.

Special days, when in effect, force all schedules that have time elements assigned in the special day schedule to adhere to that special schedule. If a weekly schedule has no elements assigned to the special day, it will remain on the normal day-of-week schedule.

Weekly Schedules.   The weekly schedules screen shall display a full week with holidays and special days included. The scheduling shall be done "spreadsheet" style, by moving the cursor to the appropriate time cell and typing in the new time. The occupancy and vacancy time cells shall be directly accessible from the operator's control language and thus assignable to virtual points of the system.

Monthly Schedule. A recommended option is to allow the choice of monthly schedules. The monthly schedules screen shall display a full month with holidays and special days included. The scheduling shall be done "spreadsheet" style and must be capable of display and editing in the same fashion as the weekly schedule. The occupancy and vacancy times must be directly accessible from the operator's control language and thus assignable to virtual points of the system.

Special Features. The schedule mechanism shall have an override feature which will override a specific time cell for a single occurrence. The cursor could be placed on top of the time element to be overridden and the override command invoked. The override time is entered and takes precedence. After a single execution, the override is deleted automatically, and control is returned to the original cell.

The schedule mechanism shall have a temporary feature. A temporary time element may be added to any schedule cell. It will be deleted automatically after a single execution.

Optimum start/stop modules shall have the current calculated start/stop times accessible through the programming language.

Modulating Controllers

Current HVAC system design practice still utilizes linear action for modulating control. This is not an energy-efficient means of control because maintaining linear action requires a significant pressure drop across the controlling device, increasing the size and power requirements of the equipment. This document encourages vendors to develop new expert controllers that will operate nonlinear as well as linear control loops with automatic tuning expert control.

At a minimum, a simple means of establishing controllers that provide proportional plus integral and derivative (PID) control is required. These controllers must have the following features:

1. All factors in PID controllers must be entered and expressed as gains, limits, etc., and in terms with physical units.

2. Actual control algorithm and its operation must be documented in the operator's manual.

3. Capacity must be provided to have different gains at different values of the proportional error signal.

4. The ranges of integral and derivative gains must be sufficiently high that certain kinds of loops can be "jogged" into setpoint by the controller.

5. Convenient integration into the control language is needed so that the use of each controller is apparent in the language code.

6. Capacity for coordination of multiple controllers for a single point and/or overrides is needed so that the language can switch controllers for an output under certain conditions or override all controllers with a specific value to the output under other conditions.

7. Capacity must exist for any or all controller factors to be system variables.

Intelligent Terminal Devices

The use of intelligent digital-controlled terminal devices has added a new and valuable dimension to energy management system effectiveness. It also creates some problems for control languages. The programming language (OCL) must be consistent at all levels of the system (see Sec. 2.1H). Furthermore, because there are often large numbers of individual devices that must be started and stopped from some central logic circuit, commands and calculations must not be made confusing by their large numbers. The following recommendations are for dealing with these large numbers of points effectively. This section is preliminary, because experience with full digital building control is too small for standards. The following features are now recommended:

Global Commands.   Commands allow wildcard characters in the point names such that an entire class of points can be commanded with a single expression.
Example:

```
START(BOXFAN*)
```

This expression would start any point whose first six name characters are BOXFAN.

```
START(BOX?FE)
```

This expression would start any point whose first three characters are BOX and whose fifth and sixth characters are FE.

DO Loops.   DO loops allow a large number of commands or calculations to be performed with a minimum of program use. DO loops should incorporate the wildcard feature also.

Example:

```
FOR I = 1, 75
    DO
            AVEST(I) = AVE(ST(I)E, ST(I)W, ST(I)N, ST(I)S)
            MAXST(I) = MAX(ST(I)E, ST(I)W, ST(I)N, ST(I)S)
            MINST(I) = MIN(ST(I)E, ST(I)W, ST(I)N, ST(I)S)
    ENDDO
```

This expression will calculate the average, maximum, and minimum of four space temperatures in each of 75 zones and place the value in 75 variables AVEST1 through AVEST75, MAXST1 through MAXST75, and MINST1 through MINST75, respectively.

Special Commands. A separate set of commands must be established to permit the transfer of programs, controllers, or other database information between levels of intelligent control. This distinction is made based on the assumption that the bus/trunk level of the terminal controls may be physically separate from the other general system bus/trunk levels.

Terminal device name: A minimum of 12 alphanumeric characters shall establish the unique address of each terminal device similar to a point name.

Special commands must be capable of being executed either from the OCL program or manually from the keyboard to manipulate the database in each terminal device. These commands are as follows:

```
LOAD(terminal device name, nnn…)
```

This command causes the whole database or a portion of it to be downloaded to the terminal controller, where nnn…is the name of a database file, a single program, controller, or other database/program module that can be named and loaded as a unit to the terminal device.

```
SAVE(terminal device name,nnn…)
```

This command causes the whole database or a portion of it to be saved to the storage media, where nnn…is the name of the database file to save to. Any single program, controller, or other database/program module that can be named and identified as a unit can be saved this way.

```
CLEAR(terminal device name)
```

This command clears the database from the controller. All outputs remain in the last commanded state when this command is issued.

```
CLEARPROG(terminal device name)
```

This command clears the OCL program from the named terminal device, but leaves intact all the point database.

Program Support Features

Program Editing. System must include a means of editing OCL programs with a full-screen editor either internally or externally. Systems should also be capable of accepting programs from ASCII files that have been prepared on other computers or word processors.

If the primary editor requires external hardware or software, it must be included in the standard system configuration to support the editor.

Note: Easy editing of both the OCL program and point names must be provided. It is not permissible to prevent database modifications because points are involved in OCL programs. If a point name is changed in the database, the automatic change of all programming occurrences of that point must be provided upon concurrence of the operator. If it is removed from the database, any program line where the name appears must show an appropriate error signal for undefined character when the program is viewed, edited, or printed.

Program Debugging. A method shall be provided of troubleshooting program errors, both syntax and runtime errors, that notifies the operator of the location of the line which is at fault. It is recommended that a trace feature also be provided to allow the operator to troubleshoot program problems. The program must provide self-diagnosis for syntax errors during editing and must provide runtime errors that include at a minimum

Endless loops

Runtime execution errors

The system must employ means to recover from OCL runtime errors without a general panel or program failure.

Program Disable. A method of disabling program sections for startup and debugging shall be provided.

Program Backup. A means of providing a rapid backup of individual programs on standard magnetic-disk media must be given so that programming changes can be saved without saving the entire database. (Complete database save features must be provided in addition to this feature.) All backup and documenting procedures must be consistent in features, time, reliability, and ease of operation with the current state of the art. Systems requiring more than a few minutes to back up the entire database are not acceptable.

Glossary

Application-Specific Controller: Digital controller that is intended to connect and operate one or more pieces of equipment that serve a specific function. Application-specific controllers usually have fixed-control algorithms with limited operator adjustment.

Automatic Communications Network: A network for distributed controllers in which all required transfers of status, value, and command information are accomplished entirely automatically.

Baud: A measurement of communication speed—bits per second of data.

Change-of-Value Update: A means of updating the value or status information from one DDC panel to another. In change-of-value update, information is transferred only when point's value or status changes by more than a predetermined amount.

Chilled Water Loop: A closed hydronic loop of water used to provide cooling.

Coil: A heat-transfer device that exchanges heat between a fluid (usually water) and air.

Communications Controller: A microprocessor that controls the communications into and out of a computer device.

Communication Trunk: A connection among computer devices that acts as a path for information to be exchanged according to a predetermined protocol.

Controller: A device that provides a signal or action which regulates the operation of one or more pieces of equipment.

DDC System: A complete control system made up of direct digital controllers linked together in a network by one or more communication trunks.

DDC System Architecture: The way in which a DDC system's components are connected to one another.

Direct Digital Controller: A controller that employs direct digital control (DDC) to regulate the operation of one or more pieces of equipment.

Distributed Control: A control system in which the control functions reside in various system devices or components.

Dynamic Control: Control strategies based on anticipated changes in conditions over time (sometimes called **feed-forward** or **anticipatory** control).

Energy Management System (EMS): The precursor of the DDC system. An EMS is a digital building controller that supplements the control of other stand-alone controllers. A DDC system is a digital controller that provides complete building system control.

Erasable Programmable Read-Only Memory (EPROM): This is nonvolatile memory that can be changed by a means that depends on its type.

Flash Memory: A type of EPROM that can be erased by applying a specific voltage.

Full DDC System: A building control system that employs direct digital controllers for all control. This term is generally used to describe DDC systems that employ DDC of the terminal boxes as well as the central equipment.

Function Block Programming: DDC programming that employs a library of preprogrammed routines (blocks) that are capable of being linked together in order to meet job-specific requirements.

Graphics-Based Programming: DDC programming that permits an operator to program with graphics symbols instead of the text characters and key words used in line programming.

High-Performance DDC System: A DDC system that is employed to provide levels of comfort and economy that are significantly better than the norm.

Heating, Ventilating, and Air Conditioning (HVAC): All equipment employed to provide comfort conditioning in a building.

HVAC Controls: The system or subsystem that is employed to control the heating, ventilating, and air conditioning system in a building.

Input/Output Devices: The devices of a control system that interface to the system(s) being controlled.

Interlock: A means of ensuring that the operation of one device is controlled or limited by the operation of another.

I/O: Input/output.

Line Programming: DDC programming that is similar to the line-programming techniques employed by the general computing industry.

Megabaud: A measurement of communication speed—millions of bits per second of data.

Operator-Interface Console: A workstation from which a DDC operator can program, monitor, or operate a DDC system.

Operator-Interface Device: A small (usually portable) device through which an operator or maintenance technician can interrogate or control the system at a direct digital controller.

Operator's Control Language (OCL): A guide for programming performance (see Appendix).

Output-Oriented: A recommended method of DDC programming that organizes all programming by the individual DDC system outputs.

PC: Personal computer.

PID Loop Controller: A proportional control-based logic block that modulates analog points, utilizing integral (control error over time) and derivative (rate-of-change) factors as well as proportional error to determine the output signal.

Pneumatic Controls: A mechanical control system that employs air as the signal medium.

Point Status: The binary result of a DDC digital input point.

Point Value: The scaled result of a DDC analog input point.

Program Module: A program or section of a program that completes a specific control or calculation task.

Read-Only Memory (ROM): Memory that can be read and used for computer operation, but cannot be changed after initial programming.

Real Time: Concurrent, dynamic.

Stand-Alone Panel (SAP): DDC monitor panel that contains all hardware, firmware, and applications software to provide continuous control of connected components based on information it collects regardless of whether it is connected to other panels or equipment.

System Points: DDC input and output connections to the system(s) being controlled and/or monitored by the DDC system.

System Variables: All digital statuses and analog variables such as enablers and setpoints that are employed by a DDC system but not physically connected to the system(s) being controlled.

Tag: Identifying label for devices and cables. In DDC applications these tags should reference the associated point name.

Terminal Regulate Air Volume (TRAV): A variable-air-volume HVAC system in which the regulation of the supply air temperature and flow is provided directly from information supplied by the terminal unit direct digital controllers.

Terminal Unit: Extended components of HVAC systems that provide the comfort conditioning throughout the building. VAV boxes, dual-duct mixing boxes, or heat pump terminals are considered terminal units.

Throughput: The amount of processing capacity per unit time of a DDC processing or transmission device.

Time-Interval Update: A means of updating the value or status information from one DDC panel to another. In time-interval update, information is transferred to requesting panel(s) at specific timed intervals regardless of a change in the point's value or status.

Tri-Services Specification: An early ill-fated attempt by the armed services to standardize emerging DDC system technologies. The inevitable failure resulted in enormous dollar and credibility costs to the services and the controls industry.

Unbundle: A method of gaining more flexible control of points operating by fixed algorithms in certain direct digital controllers. Some manufacturers of such equipment permit redefining these points in the database of supervisory controllers and directly controlling them from this higher-level controller. The process of redefining the point in the more flexible controller is called **unbundling.**

Unit Controller (UC): Small stand-alone direct digital controller with only a few input and output points capable of operating typical HVAC terminal units such as VAV boxes.

Variable-Frequency Drive: Solid-state motor contractors that have the capacity to control standard ac motors at variable speeds by changing the frequency of the ac output to the motor.

Variable Air Volume (VAV): HVAC system type that has the capacity to adjust the amount of cooling provided to each zone by adjusting the volume of air supplied to the zone.

Index

About the Author

Thomas B. Hartman, P.E., was the founder of The Hartman Company, a high-technology engineering firm specializing in applying advanced technologies to commercial and industrial energy management activities. For the past 25 years, he has been involved in developing innovative computer-based solutions to control problems that have been widely adopted in the industry. A contributing editor to *Heating/Piping/Air Conditioning* magazine, Mr. Hartman has written extensively about DDC technologies, and he is a frequent speaker on DDC systems at major industry events.